Parametric Cost Modeling for Buildings

Successful cost management and value engineering in construction is based on accurate and early estimations of cost, and this book is the quickest route to creating a cost plan from scratch. The budgeting system described in this book will help the reader to:

- document project scope at a level that provides excellent cost control at design stage;
- establish the parameters of potential sites before selecting one;
- determine the amount of financing needed before deciding to bid on a project;
- make a detailed and robust building project budget;
- determine the rental rate necessary to see if a building project will be marketable.

The technique used is a parametric cost system, not the square foot cost system used by most who quote an up-front building cost. To help calculate the parameter quantities and price them as quantified, this book comes with five electronic templates to calculate program scope: space, configuration, HVAC loads, plumbing, and electrical. It also includes:

- the author's parametric cost database and cost template to prepare the construction estimate;
- a soft cost template to price out all related program costs, convert them to a monthly cash flow, incorporate financing costs, and then reveal the final budget;
- an operation and maintenance annual cost template to calculate those variable and fixed costs necessary to run the building and then convert the result into the necessary rental rate to capitalize all costs.

The spreadsheets, data, advice, and templates are all introduced through a detailed case study, placing everything in an easy-to-understand practical context. This will prove an invaluable guide not only for estimators and cost engineers, but also developers, clients, and architects.

Donald E. Parker is an independent building consultant whose long career has included establishing the first value engineering program for the US General Services Administration, managing their nationwide cost estimating program, performing value engineering design review services for the Kingdom of Saudi Arabia, and positions as Executive Vice President of National Government Properties and Senior Vice President of RRP Corp. He is a Fellow of the Society of American Value Engineers International and President of the prestigious Lawrence D. Miles Value Foundation.

Parametric Cost Modeling for Buildings

Donald E. Parker

Routledge
Taylor & Francis Group

NEW YORK AND LONDON

First published 2014
by Routledge
2 Park Square, Milton Park, Abingdon, Oxon, OX14 4RN

and by Routledge
711 Third Avenue, New York, NY 10017

Routledge is an imprint of the Taylor & Francis Group, an informa business

British Library Cataloguing in Publication Data
A catalogue record for this book is available from the British Library

Library of Congress Cataloging-in-Publication Data

Parker, Donald E. (Donald Elmer), 1946–
  Parametric cost modeling for buildings / Donald E Parker.
    pages cm
  Includes bibliographical references and index.
  1. Building—Estimates—Data processing. 2. Building—Cost control—Mathematics. 3. Parameter estimation. I. Title.
  TH438.15.P37 2014
  692'.5—dc23
  2013044634

ISBN13: 978-1-138-01615-6 (pbk)
ISBN13: 978-1-315-77610-1 (ebk)

Typeset in Sabon
by Apex CoVantage, LLC

Printed and bound in Great Britain by
TJ International Ltd, Padstow, Cornwall

Contents

Acknowledgements

The author wishes to acknowledge the contributions made by Al Dell'Isola who, with me, co-authored our first book, *Project Budgeting for Buildings*, © 1991, Van Nostrand Reinhold (now out of print), which served as the pilot for this work. I worked with Al Dell'Isola for more than ten years in conducting numerous value engineering workshops and cost control seminars.

In 2001, the full copyright was returned to me and I automated all of the templates using Excel, added the section on operating costs and calculating rent, changed the title to *Determining Building Worth*, and published it as a CD-ROM.

Now, with the splendid assistance of Ross Finlayson, Paul Obe, and Katherine Rollins (a daughter), who are young brilliant programmers, I am able to enhance this text with the development of a spreadsheet Excel database to facilitate the preparation of the actual construction estimate.

The editorial assistance of my wife, Mary Frances and my other daughter Ann Mutersbaugh, who is an architect, round out my support team for which I am grateful.

Preface

This text is for those developers, owners, and project managers who must quote a rental rate for a building that is not yet designed. It is for those who need to know how much financing to seek before going ahead with a project. It is for those who need to know what a program of building requirements should cost in order to determine value engineering savings opportunity.

Parametric Cost Modeling for Buildings is based on the author's 35 plus years of cost engineering experience. For the past decade the author was directly involved in the asset management of three large portfolios of commercial properties located throughout the United States—some 560 buildings, 14.6 million square feet, valued at more than $1 billion.

His early career was with the federal government where he first developed and honed the parametric estimating tools presented in this text. With the government he was responsible for the prospectus budgeting of many major building projects while he served as Director of the General Service Administration (GSA) Cost Management Division.

With GSA he was instrumental in introducing their Capitalized Income Approach to Project Budgeting and wrote handbooks to introduce the UNIFORMAT system of estimating to the agency. For more than two decades he has held credentials both as a certified cost engineer and a certified value specialist—a combination that only a handful of people possess.

The author has spent his career either developing project budgets, or working with established budgets in efforts to optimize total building costs. Also, he was directly involved with the preparation of bids to quote rental rates or total costs (which were then capitalized) required to design, finance, construct, operate and maintain large government projects over specified lifecycle periods. Much of the detailed information in this book is derived from the experience cited above plus several cost guideline manuals developed for facility directors of large building programs and from conducting numerous seminars on project cost control and value engineering.

In response to a solicitation for offers (SFO) to provide leased space for a tenant, the developer's challenge is to quote a final rental rate that is good for 10–20 years. Not only must this be done prior to starting design work, the rental rates quoted must often include sufficient cost to provide a full serviced facility (utilities, operations, maintenance, reserves for replacement management and insurance). Once this bid commitment is made and the lease is awarded to the developer the need to control cost becomes acute.

This book outlines an organized approach to determine a reasonable budget for commercial real estate that will improve the accuracy of the budget as well as enhance its use as a project management tool for overall project cost control. The book is designed for use by developers, owners, and their engineers, planners, project managers, architects and construction managers, design-build contractors, estimators, and value engineers. Its content has special significance for owners, agencies, or corporations embarking on large, new, or renovated building projects who desire an increased degree of overall confidence in their budget with better project cost control. Through its use a budget can be developed that will limit discrepancies in communications between the developer/owner and the designer. As such, it provides the basis of estimating (BOE) for control of work through the development process.

The book starts by introducing the traditional approach to budget estimating, describes the steps for a new approach, and provides an overview of the specific additional areas in which the technique can be used.

Subsequently, the book outlines the approaches recommended for developing the required input data for a comprehensive budget. Each chapter follows a step-by-step approach to preparing a budget that concludes in a rental rate necessary to support financing and developing the facility.

Accompanying this book are Excel formats, templates, cost database, and other tools illustrated in this book to assist in making the budget as described herein. They may be downloaded from the publisher's website, www.routledge.com/9781138016156, and are contained in a "zip" file that can be opened on the reader's hard drive and used.

Once the construction estimate is finished, provided Excel formats can be used to prepare the annual operating and maintenance costs for the facility, the soft cost budget for development, the cash flow schedule for computing interest during development, and the rent model for converting these costs to a rental rate.

The resulting formats from these computations provide highly organized, detailed budget documentation, which is useful to minimize discrepancies in owner intentions, design criteria, and resulting cost implications.

Donald E. Parker, PE, CCE, CVS

1 The Method and its Usefulness

Introduction

Estimating is the process of predicting (or forecasting) within acceptable variances what the actual cost will be when a given project is completed. This text focuses on preparation of the initial estimate, the budget for commercial real estate development that determines the economic worth of the project in terms of the rental rate that must be obtained to retire its debt. Once a budget for a project is established, the task then is to control cost to stay within the budget to make the prediction come true.

One evaluates a budget as either being too high, too low or within an acceptable range. An acceptable budget would provide neither too little nor too much money to do the job. A budget that is too low manifests itself in cost overruns, the cutting back of requirements, and management headaches. A high budget is just as bad; it makes a developer noncompetitive, increases the cost of financing, and reduces the margin of profit at which the project can be sold.

The budget process described in this book reduces the risk of having a bad budget because the method used is directly based upon the project's specific criteria and scope rather than being general in nature. This text, beginning with Chapter 3, is a step-by-step estimating procedure that walks the fine line between using too little data to prepare a budget and requiring too much data to prepare a budget.

Conventional Practice

The most common method of budget preparation for commercial real estate (a new office building, hotel, shopping mall, etc.) is to estimate it on a cost per square foot basis. A survey taken by the Veteran's Administration[1] in 1974, which the author still believes is valid, indicated that the square foot method of estimating was used by 82 percent of all architect-engineer (A-E) firms to prepare budget estimates.

These budgets, when compared to the actual construction low bid for the projects, showed the following ranges:

- extreme deviation   (28 percent above low bid, 38 percent below range = 66 percent   low bid).
- mean deviation   (13 percent above low bid, 16 percent below range = 29 percent   low bid).

Approximately 12 percent of the A-E firms surveyed used a modular quantity take-off method for budget preparation. This method indicated some improvement in accuracy over the square foot method of budget estimating. When compared to bid results, the deviations were as follows:

- extreme deviation   (21 percent above low bid, 10 percent below range = 31 percent   low bid).
- median deviation   (14 percent above low bid, 10 percent below range = 24 percent   low bid).

One of the largest variables in budgeting is the capability of the firm to exercise effective cost control through design development. From the above data one can see that cost control using a square-foot budget as a basis is virtually impossible. The ability to control cost to a budget seemed to improve as definition of the budget basis improved. As seen from the survey, the most commonly used budget technique for facilities is the use of the following overall method:

- identify the type of facility;
- budget the dollar per gross square foot ($/gsf).

The minimum information necessary for this type of budget is to know:

- historical cost for the facility type;
- desired gross square footage;
- geographical location;
- desired completion date.

   Too often this minimum information is all that is known or used when budgets are prepared. Project budgets developed on this basis are totally inadequate for controlling cost during future design stages, and this method does not provide confidence that it includes all project required scope and criteria without overstating cost. For example, construction budgeting publications show a wide variation in historical cost per gross square foot, depending on the type of building.

   Within building types, cost ranges similar to that shown by the following sample data[2] are typical:

- offices—mid rise $105–$172/gsf
- parking garages $37–$93/gsf
- auditoriums $111–$219/gsf
- court houses $167–$271/gsf.

Budgeting solely on this basis is "pick a number." When budgeting is performed in this manner one is limiting or selecting, without documentation, factors such as facility quality level, program content, space efficiency, facility configuration, and future lifecycle cost experience. Because these are undocumented, they cannot be controlled against the budget.

Budgeting Objectives

The developer really has two objectives when a project budget is developed: to win the bid and/or to secure funding. After these objectives are accomplished the need to control cost becomes paramount.

Win Bid and/or Secure Funding

The developer must win the bid with the lowest rental rate but not at the expense of failing to secure funding. These two objectives go hand in hand. If the rental rate is too low then finding permanent financing to take out the construction loan will be more difficult. The long term lender is interested in the estimated revenue stream to pay back the debt and operating expenses before equity.

Most developer and owner cost and cost control problems are created at the budget-planning stage of a project. More often than not, owner needs are not fully known and thus are oversimplified intentionally or unintentionally. Or, even worse, client needs are understated in order to win the bid or justify the project. For speculative development the specific needs of the client tenant are unknown.

Beyond responding to a request for proposals, a project starts in many ways; an idea in an executive's mind, a scheme from an advanced planning group, a request from a sales department for more product, a changing profit picture, or a need for more space.

The project budget is often prepared or used by marketing personnel to perform economic analysis to determine a return on investment (ROI). A lower budget provides a better ROI.

It is easy to over-simplify client needs at the budget stage and do it quite innocently without a definitive building concept. The owner/developer is trying to quantify a dream. Yet, it is known that the reliability of a budget improves in proportion to the amount of information available when it is created. The opinion that estimating is an art and not a science is only partially correct. Only when there is no information is estimating all art.

The estimating method described in this book indicates where information that can be used is available and how to develop useful information from project requirements.

Cost Control

The control cost objective becomes a requirement once funding has been received. Now that management is committed to a fixed rent and/or a fixed price, everyone must achieve it.

There is a difference between managing cost and controlling cost. To manage something is to succeed in accomplishing. To manage by cost is, then, to succeed in accomplishing a cost objective. Management is the act or manner of handling, directing or controlling something. Control is a process, that is, a systematic series of actions directed toward some end.

The dictionary[3] defines the term *control* in two ways: (1) to check or verify by comparison with a duplicate register or standard; (2) to regulate, exercise authority over, direct or command to take corrective action. This definition of control, when coupled with the term *cost,* gives no indication or solace that costs would not rise if cost control were practiced. Cost control does not promise the end to the problems of management, be they inflation, overdesign, or anything else.

What it indicates is that one must have a budget baseline against which to compare so that management can spot deviations in time to take corrective action. The strong assumption in the term *control* is that management is willing to exercise authority—to make a decision.

Therefore, it is important that the method used to develop the project budget be precise enough to provide a basis for monitoring throughout the detailed design process. The estimating system proposed by this book does the job because it is based upon determining design parameters and quality levels, then pricing on a conceptual basis in enough detail to allow the control process to be effective. If the budget used to seek the project financing cannot be used in this fashion, then control of the budget during execution will be difficult or impossible to achieve.

The problem lies with the fact that many feel that cost control means the control of money or a budget review. In fact, when cost control is mentioned the first thing they do is look for the estimate to see what prices can be cut. Those that control costs by looking solely at estimates, money, or cash flow are overlooking key factors. One controls cost by controlling scope, not dollars. The key to achieving cost control through scope control lies in the definition of scope (see Chapter 2). The old-fashioned idea of viewing scope as building square feet is not sufficient. Scope control is achieved by identifying all requirements and generating a baseline document to record them. Such a system requires close monitoring by management, but it does permit verification to take place in order to regulate, thereby achieving the control function.

Response to an SFO

The estimating method proposed by this book has been used by the author many times to respond to a Solicitation for Offers (SFO) for the development of a new office building, clinic, computer laboratory, and a wide variety of commercial projects. SFOs are the common way federal and state governmental agencies request proposals for building space. The SFO process has also been used by major corporations to solicit space.

Offerors submitting proposals in response to SFOs could be offering to lease existing buildings or to provide a new building for lease. SFOs normally provide the type of space required, net square footage, and requirements for structural, plumbing, electrical, HVAC, elevators, and special systems. They often also include requirements for janitorial, utilities, maintenance, and operation.

Existing Buildings

An effective way to offer an existing building is to compute the building systems as if they were new, with requirements taken from the SFO. Use the method described below for new buildings with the following caveat—restrict your computations to use the existing buildings' square footage and shape. Then compare each system with the respective systems in the existing building to identify deficiencies that need to be corrected to respond to the procurement.

In this manner a capital improvement budget estimate can be developed for each system that can be amortized in the proposed rental rate over the term of the lease.

New Buildings

The estimating process for budgeting new buildings as described in this book involves the following steps:

Step 1—Compute space requirements (Chapter 3).

- list net areas;
- convert to gross area;
- calculate support space.

Step 2—Establish configuration (Chapter 4).

- determine number of stories;
- document shape (length, width, height);
- compute site area.

Step 3—Determine architectural/structural quantities (Chapter 5).
Step 4—Determine mechanical/electrical quantities (Chapter 6).

- make HVAC model;
- make plumbing model;
- make electrical model.

Step 5—Prepare construction cost estimate (Chapter 7).

• use systems parameter cost template with database.

Step 6—Determine soft costs (Chapter 8).

• use spreadsheet software.

Step 7—Compute operating cost and rent (Chapter 9).

• use spreadsheet software.

After reading the SFO, the above process takes the author approximately four hours to go through the seven steps to prepare a budget. Once the first budget is prepared it is easy to perform "what if" scenarios such as:

• What if the building were eight stories in lieu of ten stories?
• What if the exterior wall had a thermal U factor of 0.32 in lieu of 0.10?

These are easy questions to answer using the author's system. They are extremely difficult to answer using the cost per square foot system.

Determining Site Objectives

Budgeting is easier if the developer has a known site. One can tell at Step 2 of the process if the site is large enough to accommodate the project within zoning restrictions or if it is too large (and therefore might contribute more cost than necessary to meet minimum requirements).

If a site, however, does not exist the developer will need to search for alternative sites and take an option on one to respond to the SFO. Assuming site selection meets the client's needs for location, transportation, amenities, etc.—the question then becomes: how large?

Step 2 of the process described above allows one to compute alternative building configurations that meet program requirements. The objective is to determine the ground area (footprint area) of the building. This varies depending on the number of stories. Site size that is needed is determined by adding the following ingredients to meet zoning requirements:

- building footprint area;
- parking area;
- landscaped area;
- setback area.

Before a site is optioned or acquired, lenders normally require the following data to ensure reduced cost risk. It is a good idea for a developer to consider these in selecting a site to offer to see if they have any impact on the cost being budgeted for amortization in the rental rate:

- environmental study (Phase 1);
- soil borings;
- utility "will serve" letters;
- title—easements and survey;
- code and zoning analysis;
- preliminary site layout.

Alternate costs for the site as well as alternate impacts on construction cost can easily be placed into the spreadsheet at Step 6 and revised rental rates computed in Step 7.

Analyzing the Price to Pay for a Site

On occasions the developer will find a site in a community where its development rights are determined by local ordinance. Owners will set a price for those development rights and appraisers will determine site value. The budgeting method described in this book can be used to provide an alternative check on the reasonableness of the price to be paid.

For example, the author was presented with a site that had 250,000 gross square foot development rights in the city. The question was, what is the value of the land, what price should be offered? Here are the steps to answer that question:

1. Estimate the construction cost using Steps 1–5.
2. Estimate the soft costs (without land) using Step 6.

3. Compute the net rentable area (assuming a 15 percent core factor, multiply gross area by 0.85).
4. Multiply the market rental rate by the net rentable area to determine potential gross income.
5. Subtract operating costs (Step 7) from gross income to determine net operating income.
6. Subtract total project cost from net operating income to determine value of the site.

Now you can judge if the asking price is reasonable or what adjustments to rent or other cost factors might have to be made if you purchased the site at its asking price.

Determining Price of an Existing Building

There are three appraisal approaches to determining the price or value of an existing building. These are the market approach, income approach, and replacement cost approach.

Steps 1–6 of this text can be used to determine the replacement cost of the building. All one needs to do is make sure that the data in Steps 1 and 2 imitate or replicate the existing overall measurements and shape of the building.

For a project of any size it is much better to take the time to estimate the replacement cost at the parametric level of detail suggested by the author than go by instinct or "gut feeling" of an appraiser cost per square foot.

Value Engineering Design Concepts

Performing a value engineering (VE) study at the design concept or 35 percent design stage of a project has long been known to be the most effective time to realize maximum savings. In fact most system designs are locked in after this stage and changes to achieve savings must be really substantial to offset the cost of redesign.

When the A-E delivers the plans, specifications, and cost estimate for VE study the chore is to synthesize the data to isolate potential areas of opportunity to reduce cost without sacrifice of program requirements or quality. Most practitioners of VE attempt to do this by preparing a "cost model" from the A-E estimate with a corresponding "worth model" using intuition and judgment.

The author recommends reverse engineering the A-E design to determine what it "should cost" using the estimating method described in this book. Take the same program requirements data and design criteria set forth in the SFO and make your own budget estimate following Steps 1–5.

In Step 2 you can make alternate configurations to develop alternate parametric quantities. You can also model the existing design by inputting the number of stories just as you would an existing building. All of this will result in your own computations of gross space, building envelope, HVAC, plumbing, and electrical quantities. If you price out the quantities you develop for achieving the client's program on a system basis you can compare them to the A-E's cost.

This process allows one to check out space and system efficiencies in terms of parametric quantities. This often provides an additional benefit to challenging system materials, i.e. precast vs. brick, VAV system vs. FCU system, etc.

Designing to Budget

The task of holding project costs at the level initially accepted by the developer/owner depends completely on a team effort, an effort identified by the term *project cost control*. The project cost control team members are the developer, the project manager, the cost engineer, the design professionals, and the owner's representative.

The task of project cost control is to ensure that the objectives encompassed in the project budget, as defined in terms of quality and quantity, are achieved in the final project. Only rigid discipline in controlling design development makes accurate budget estimating possible.

To achieve the minimum discipline necessary to control costs, the budget estimating system provided in this book includes:

1. establishing uniform procedures and formats for use by owner and all members of the building team;
2. performing certain basic design programming and engineering before the budget is prepared;
3. using the budget documentation to control project scope between owner and designer.

Each of these requirements permits "verification" to take place in order to "regulate," thereby achieving the "control" function.

Summary

Proper budget preparation can provide better information to management on which to base investment decisions. And, once the investment decision is made, the budget can be used by management as a vehicle to control project scope and design decisions in advance of experiencing a cost overrun.

If the purpose of project budget estimating is to establish the amount needed for a loan or an appropriation to meet a given expectation, then the method used to develop the budget should aid in ensuring that one can control costs to stay within the budget. Thus, to control design, basic design parameters and quality levels must be established during budgeting. They must then be used as guidelines in developing the ultimate project design.

Notes

1. *Ingredients for Accurate Construction Cost Estimating*, by G.M. Hollander, Actual Specifying Engineer, June 26, 1974.
2. R.S. Means, *Building Construction Cost Data*, 2013, Norwell, MA: Reed Construction Data LLC.
3. *Webster's New Collegiate Dictionary*, 1975, Springfield, MA: Merriam-Webster Inc.

2 Fixing Project Scope

Introduction

The dictionary defines scope as "the range or extent of a concept, the room or opportunity for freedom of action." Scope is defined by words, drawings, and cost figures.

For development, design-build or turnkey organizations these documents result in the preparation of a proposal to be offered to a client. For an operating organization, these documents are submitted to management for approval, appropriation, or financing. Subsequently, they go to procurement for design contracting.

In a standard prepared for the AACE International[1] by the author, facility project scope consists of three broad elements: performance, schedule, and cost requirements. To most architects, scope consists merely of the owner's program needs for net square feet of space.

If square feet is all that is specified, that leaves a wide range of opportunity for freedom of choice of everything else in the project. With such maneuvering room, cost will also have a wide variance.

Effective control of cost through scope is limited to the level of detail provided in the baseline used to prepare the budget.

Budget Categories

At this point, it is well to define those budget elements developers and owners attempt to control. Figure 2.1 illustrates the five budget categories used by the author to compute total budgeted cost for a project in response to a GSA solicitation for offers (SFO). However, most of these costs occur in all projects, for both government and private sector.

	Total Project Cost (ETPC)			
Estimated Construction Cost (ECC)	**Estimated Site Cost (ESC)**	**Estimated Design Cost (EDC)**	**Estimated Management Cost (EMC)**	**Estimated Financing Cost (EFC)**
Base Building	Land	A-E Fee	Bid Proposal	Financing Fee
Unit Price Items	Environmental	Reimbursables	CM Fee	Interest
Reimbursables	ALTA Survey	Layout	Local Fees	Interest Reserve
Bond	Soil Investigation	Special Studies	Permits	
Insurance	Title Work	Design Review	Testing	
Demolition	Appraisal	VE Services	Developer Overhead	
Contingency	Taxes	Observation	Reimbursables	
	Code/Zoning	As-Built	Legal	
	Analysis	Operating	Insurance	

Figure 2.1 Program budget categories. Total Project Cost = ECC + ESC + EDC + EMC + EFC

For certain projects the author worked on overseas, items for furnishings, equipment, and import duty was added. On projects where the developer does not have leases in place, one would need to add items for lease advertising and commissions.

This text focuses on the control of the Estimated Construction Cost (ECC) budget element. The focus is on construction cost because that element represents 50–75 percent of the total acquisition cost for a facility. And construction cost, to some extent, influences the amount budgeted in the other elements. ECC is also the budget element that receives high owner and public visibility at bid opening and is most often the ultimate judge of cost control achievement.

However, equal time and effort should be devoted to controlling the other budget elements shown in Figure 2.1 (ESC, EDC, EMC, and EFC). Achieving construction within the construction budget but overrunning in design, site, financing, or other management cost areas can make a project unattractive. The principles set forth in this text for documenting and controlling ECC apply equally as well to these areas of cost.

Additionally, while not the subject of this book, the operation and maintenance costs estimated for the project can be controlled using the same techniques.

Design Programming

In the last decade a new professional service has evolved called design programming. Design programming is an independent service that quantifies and defines facility needs. It is an activity that assists in budget setting and precedes the contracting for architect-engineer design services. Because the purposes of design programming and facility design are distinct from one another but are interrelated, most organizations will retain these two services from different professional firms or perform design programming with their in-house staff.

Design programmers often refer to project performance requirements as the design basis as depicted in Figure 2.2. The schedule

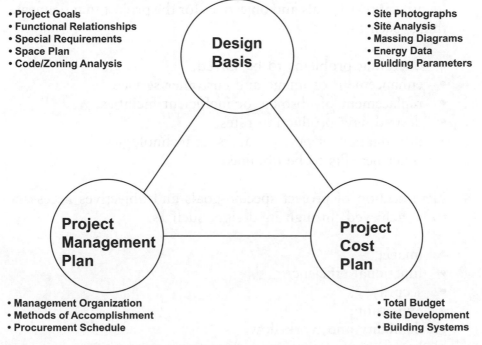

Figure 2.2 Three elements of scope

portion of scope is expanded and referred to as the project management plan in design programming. The budget becomes the cost plan.

Generating Artificial Intelligence

It is to be expected that budget estimates will be prepared at a time when developer and owner requirements for the facility are still unresolved in significant detail. Yet there is much good information available that can be used if one has some experience and knows specifically what needs to be obtained.

To start, one should attempt to learn, and document, the following information about the project:

1. A statement of need for the facility in terms of the developer or owner's overall market or business strategy.
2. Identification of developer goals and objectives such as:

    *   investment or sale of property objectives;
    *   desired rates of return.

3. Identification of goals and objectives for the project that delineate things such as:

    *   operating problems to be solved;
    *   enhancement of image and customer service;
    *   replacement of obsolete or inefficient facilities;
    *   desired new production rates;
    *   introduction of new products or technologies;
    *   other benefits to be obtained.

4. Identification of project specific goals and objectives necessary to be achieved through its design, such as:

    *   image;
    *   functional efficiency;
    *   security;
    *   expansion;
    *   flexibility and work flow;
    *   human performance and amenities;

- energy conservation;
- technical performance of its systems.

Next, one needs to define requirements in terms of the seven key cost drivers discussed later in this chapter. However, even under the worst-case condition of no information, one can proceed with making a proper budget by making assumptions regarding the seven key cost drivers and using previous knowledge of:

- tenant or owner standards and criteria;
- national and local building codes;
- accepted professional practices.

Using knowledge of past performance standards, codes, criteria, and quality levels, one can base prices upon assumptions of materials and production techniques. If schedule is assumed as well, one can assume production rates based upon each production technique, the crew sizes, the need for overtime, and the need for escalation.

The resulting budget must be the cost engineer's interpretation of the developer/owner's requirements supplemented by intelligence from the above sources and past practical experience.

No one should dispute the accuracy of a cost estimate made on this basis. One should review the accuracy of the assumptions upon which it is based. Therefore, it is the responsibility of the cost engineer, working with whatever members of the developer/owner's building team he can find, to:

1. Interpret what is required and then forecast the expected cost of supplying those requirements.
2. Fill in missing information with the assumptions used for including items, excluding items, quantification, item description, and pricing.
3. Supply a "concept" of the project requirements and its ultimate design solution through representation in the budget estimate.
4. Leave a traceable path identifying the assumptions upon which each estimate has been structured.
5. Provide a budget for each concept that, if financed, could be achieved if adhered to during design.

All of this becomes the basis of estimating (the "BOE") for control of the project through design and construction to control cost.

Budget Estimating Standards

Adherence to good estimating standards for budget preparation will help to ensure achievement of developer/owner objectives to secure appropriate funding and control cost. The standards that should be met in budget estimating include balance, completeness, accuracy, and functional separation.

Balance

Between the various disciplines and building systems making up the project, the estimating effort submitted for each discipline should be proportional in relationship to the total cost. Balance of estimating effort is achieved by quantifying each of the building systems making up the project. Lump sum amounts should be held to an absolute minimum.

Completeness

For the budget to be acceptable, it must include all relevant budget elements and all elements of proposed project work. Completeness is achieved by following the UNIFORMAT system code of accounts, thus ensuring that no system is accidentally omitted.

Accuracy

The hallmark of a quality estimate is accuracy, and without this element it loses its credibility. Computations should accurately reflect the concept being budgeted. While accuracy is desired, too much precision in presenting budget figures can be misleading and can be more work than is warranted for the level of information and assumptions being made at the budget preparation stage. Therefore, it is appropriate to round off individual budget items to the nearest $100 or $1,000 depending on the project magnitude.

Functional Separation

Separate estimates should be made for each individual building structure within a project. If one project or site includes more than one structure, separate estimates of each are necessary. To lump the cost, for example, of the exterior wall system of three structures into one budget number defeats the purpose of using the budget for cost control.

Key Cost Drivers

Adequately determining scope is achieved by defining the project's technical content and associated quality levels and quantities. The tools available to do this are requirements in solicitation for offers (SFOs), precedence, standards, criteria, codes, and guide specifications.

In addition supplementary documentation is often available to assist in defining scope in the form of surveys, drawings, models, photographs, and reports. To a cost engineer, there are seven broad areas that, when established, are key in determining project cost for any type of facility. These are:

1. Functional Areas

The net square feet of each type of space to be provided in the project should be listed by type. The sum of all this space should represent the proposed tenant requirements for the facility.

Knowledge of these quantities of space facilitates the budgeting of equipment, finishes, and various system quantities (e.g., power, lighting, heating, air conditioning, plumbing, and ventilation) for each space type.

Chapter 3 explains how to convert functional area requirements into gross square feet for construction.

2. Occupancy

Many features of a facility are dependent upon the number of occupants designed to use it as well as the operating profile of the facility. The following items should be documented:

- permanent employees: executives, supervisors, workers by type;
- part-time employees by type;
- number of visitors related to the space types within the project such as:
  - auditorium/theater;
  - training;

- ○ shopping;
- ○ courtroom;
- ○ library;

- operating hours;
- number of shifts;
- number of employees by shift.

Knowledge of people using the facility is important because this data influences the amount of plumbing necessary, the amount of circulation for stairwells and exits, the amount of parking necessary to meet local zoning, and the sizing of support space such as lunch rooms and auditoriums.

The type of functional space planned for a facility will also determine the number of visitors it will draw. For example, space to accommodate tour groups, shopping, theater, training, and large conference facilities can be expected to increase building system requirements at a higher budget than if they were not provided.

3. Configuration

Configuration data does not mean designing the building. It does mean indicating the number of floors, height, perimeter, and volume. To achieve this also requires some code and zoning analysis for a specific site or the identification of site requirements from the configuration assumed. Chapter 4 provides guidance in determining this data.

4. Design Parameters

Once a program and configuration are established, one can calculate the design parameters for the major systems of a process facility or building. The parameter quantity for each system is dependent on the criteria used or assumed.

Generally, there are four major systems that are key cost drivers dependent on engineering calculations based on criteria. These are the structural system, mechanical system, plumbing system, and electrical system.

Computation of parameter quantities is explained in Chapter 5 for architectural and structural systems and Chapter 6 for mechanical and electrical systems.

5. Special Systems

This key cost driver involves the identification and quantification of all special systems and features to be provided. Normally, the decision to include them is a "yes–no" decision of the developer or owner who must anticipate or respond to the needs of the tenant.

Some of the more common special systems and features, with their normal method of measurement (where appropriate), are listed as follows:

Sprinkler—HEADS	Intercom—STATIONS
Fire Alarm—STATIONS	Standpipe—STATIONS
Emergency Power—KW	UPS systems—KW
Cranes & Hoists—EACH	Dual Fuel—GALLONS
Public Address—STATION	Security System—DEVICES
Historic Preservation—SF	Master Clock—STATION
Lightning Protection—RODS	Grounding—POINTS
Telephone—STATION	Local Area Network—STATION

6. Geographical Location

Geographical location provides essential data for use in development of project scope. It provides structural criteria (seismic and wind loading) and mechanical criteria (outside winter and summer design temperatures). It is important for determining necessary index adjustments to labor, material, and equipment costs. And, if the location is overseas, foreign exchange rates, customs, and duties can be determined.

System costs known for one location can be indexed to another location and, if the location is remote, budget elements can be added for transportation of materials and labor per diem for labor personnel. Figure 2.3 provides the listing of geographical data available for a sample project location.

DATA ELEMENT	VALUE	MEASURE	SOURCE
LOCATION			
Atlanta, Georgia			
Latitude	33	Degrees	P-89mod.pdf
STRUCTURAL DATA			
Seismic zone	2	Zone	Figure 5-2
Wind loading	100	Miles/hour	R.S. Means
Prevailing wind direction	ENE		P-89mod.pdf
Frost depth	6	Inches	R.S. Means
Snow loading	10	Psf	R.S. Means
HVAC DATA			
Winter design temperature	22	Degrees F	P-89mod.pdf
Summer design temperature	92	Degrees F	P-89mod.pdf
Humidity ratio	0.016		Figure 6-6
Solar loading			
Roof 55 Btuh/sf			Figure 6-6
North exposure	37	Btuh/sf	Figure 6-6
South exposure	111	Btuh/sf	Figure 6-6
East exposure	219	Btuh/sf	Figure 6-6
West exposure	219	Btuh/sf	Figure 6-6
Degree days—heating	3095		P-89mod.pdf
Degree days—cooling	1589		P-89mod.pdf
PLUMBING DATA			
Rainfall, 10-year, 1 hour	2.6	Inches/hr	
ECONOMIC DATA			
Construction index	88.7		R.S. Means

Figure 2.3 Geographical information

The author's "zip" file contains a file of worldwide weather data (P-89mod.pdf)[2] based upon Defense Department research that is used for the HVAC computations.

7. Schedule

Key milestone dates must be fixed or assumed to provide the scheduling data for a controllable budget. Guidance in making this scheduling is provided in Chapter 8.

Summary

As discussed above, a design programming effort should provide much of the above seven categories of key cost driver information needed to prepare a realistic project budget.

If preliminary engineering or design programming of the facility has not been performed, the cost engineer must be sensitive to the missing data and substitute assumptions that can be reviewed by management at a later time.

Notes

1. Defining Project Scope for Effective Cost Control, *AACE Cost Engineer's Notebook*, Volume 2, 2008, Morgantown, WV: AACE International.
2. Facility Design and Planning, *Engineering Weather Data*, 1 July 1978, Departments of the Air Force, the Army, and the Navy. Known as "P-89," a copy of which is provided in the .zip computer files accompanying this book.

3 Computing Space Requirements

Introduction

The first step in the estimating process is to convert the desired program requirements for space into gross square feet (GSF). This may seem to be a simple task but it grows much more complicated because of the many types of measurement systems that exist and are used by owners, contractors, tenants, and architects. It is like a stew. One must understand the ingredients to make it. And one should understand the basic ingredients of each of the following types of space in order to convert square footages from one type to the other:

- NSF—net square feet;
- NUSF—net usable square feet;
- RSF—rentable square feet;
- IGSF—interior gross square feet;
- GSF—gross square feet.

In addition, developers must contend with the following specific types of square footage mentioned in GSA leases:

- OSF—occupiable square feet;
- BRSF—BOMA rentable square feet;
- BOU—BOMA office usable square feet.

One can study the nuances between each of these types of space but for budgeting purposes it is not worth the effort. This chapter will provide some basic definitions for each type of space measurement and, to estimate at the budgeting stage, recommends simplifying the types to just three:

- NSF—net square feet where NSF, NUSF, BOU, OSF are equal;
- RSF—rentable square feet where RSF and BRSF are equal;
- GSF—gross square feet is the area constructed by the contractor and used to determine cost on a $/GSF basis, not IGSF.

Space Efficiency

The tenant's needs for space are normally programmed in terms of net square feet (NSF). However, the scope of construction must be programmed and budgeted in terms of gross square feet (GSF).

The ratio of NSF/GSF is referred to as the space efficiency. The efficiency of space varies widely, depending upon the type of space provided. Figure 3.1 provides a list of typical efficiency ratios from two sources for comparison with a suggested range to use in budgeting. These have been derived from past experience and

EFFICIENCY RATIOS

Space Type	(OSF) GSA	(NSF) Means	Budget Range
Computer area	.61		.60 – .65
Auditorium	.70	.70	.65 – .75
Cafeteria	.67		.65 – .75
Restaurant		.70	
Dining Hall			
Classroom	.66	.66	.65 – .70
Court	.65	.61	.60 – .65
Laboratory	.58	.58	.55 – .60
Library	.76	.76	.75 – .80
Medical	.55	.55	.50 – .60
Storage	.80		.75 – .85
Office—open plan	.75		.70 – .80
Office—private	.70	.75	
Parking—stacked	.95		.90 – .95
Parking—single	.90		
Garage		.85	.85 – .95
Warehouse	.95	.93	.90 – .95

Figure 3.1 Space efficiency ratios

analysis. It is important to understand the definitions of various types of space before applying these ratios.

Occupiable Square Feet (OSF)

OSF is a term derived by the federal government (GSA) for classifying space. OSF includes only that space for which the government charges rent. They do not charge agencies for public corridors or public toilets and other supporting spaces. To be specific, the following is quoted from one of their standard solicitations:

b) Occupiable Space is determined as follows:

1) If the space is on a single tenancy floor, compute the inside gross area by measuring between the inside finish of the perman-ent exterior building walls or from the face of the convectors (pipes or other wall-hung fixtures) if the convector occupies at least 50 percent of the length of the exterior walls.
2) If the space is on a multiple tenancy floor, measure from the exterior building walls as above and to the room side finish of the fixed corridor and shaft walls and/or the center of tenant-separating partitions.

In all measurements, make no deductions for columns and projections enclosing the structural elements of the building and deduct the following from the gross area including their enclosing walls:

i)   toilets and lounges;
ii) stairwells;
iii) elevators and escalator shafts;
iv) building equipment and service areas;
v)   entrance and elevator lobbies;
vi) stacks and shafts; and
vii) corridors in place or required by local codes and ordinances and/or required by GSA to provide an acceptable level of safety and/or to provide access to all essential building elements. (Corridors deducted to determine occupiable space may or may not be separated by ceiling high partitions.)

Net Square Feet (NSF)

NSF is a term normally used in the private sector. Be careful when reviewing design programs from clients for new facilities. They often mix various types of space in their requirements in order to be sure that they are provided. For example, the author has seen programs that specify:

- certain quantities of space plus circulation space;
- loading docks, telephone rooms, janitor closets;
- lobby sizes, corridor widths, etc.

GSF includes all enclosed floors of the building, basements, mechanical equipment floors, and penthouses. The rule of thumb is, when you think you can collect rent for the space because it is a special requirement (not a requirement of the building), the space is "net" space or "rentable" space. The author calls all net space "usable" space—usable and required by the tenant.

Gross Square Feet (GSF)

The rules of measurement of GSF are provided in the next section. For now, it is sufficient to say that GSF is the sum of the construction area of a building. Floors are measured to the outside finished surface of permanent exterior building walls. GSF includes all enclosed floors of the building, basements, mechanical equipment floors, and penthouses.

Rules of Measurement

The ANSI standard[1] is used by the Building Owners and Managers Association (BOMA) to measure usable area and rentable area as follows.

Usable Area

This method measures the actual occupiable area of a floor or an office suite and is of prime interest to a tenant in evaluating the space offered by a landlord and in allocating the space required to house personnel and furniture. The amount of usable floor area on a multi-tenant floor can vary over the life of a building as corridors expand and contract and as floors are remodeled.

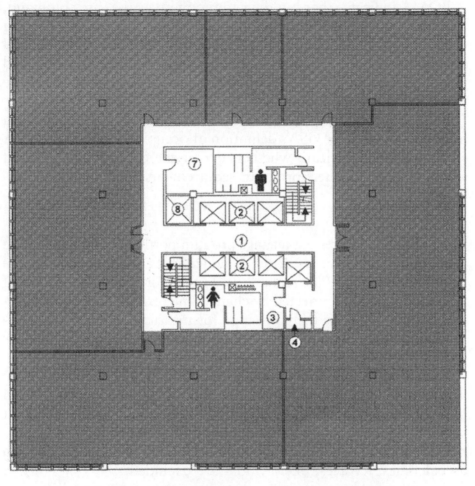

1. Lobby	2. Elevator	3. Electrical Room
4. Janitor	7. Fan Room	8. Ventilation Shaft

Figure 3.2 Illustration of floor usable area for a typical upper-level floor

Figure 3.2 illustrates usable area. Usable area is computed by measuring the area enclosed between the finished surface of the office area side of corridors and the dominant portion and/or major vertical penetrations. No deduction is made for columns and projections necessary to the building. Where alcoves, recessed entrances or similar deviations from the corridor line are present, floor usable area is computed as if the deviation were not present.

BOMA indicates that building common areas (i.e. fire command, building maintenance office, trash dumpster area, loading dock, vending areas, exercise club, retail service corridors, and security) are considered to be part of floor usable area. To avoid misinterpretation, BOMA advises that this figure should not be used without the complete ANSI document.

For budgeting purposes, the BOMA definition of usable area most closely matches GSA's definition of occupiable area. Each GSA solicitation will indicate whether any of the common areas listed above are part of their program requirements.

Rentable Area

This method measures the tenant's pro rata portion of the entire office floor, excluding elements of the building that penetrate through the floor to areas below. The rentable area of a floor is fixed for the life of a building and is not affected by changes in corridor sizes or configuration. This method is therefore used for measuring the total income-producing area of a building and for use in computing the tenant's pro rata share of a building for purposes of rent escalation.

Figure 3.3 depicts rentable area. BOMA states that floor rentable area means the result of subtracting from the gross measured area of a floor (GSF), the area of the major vertical penetrations on that same floor. No deduction is made for columns and projections necessary to the building. Spaces outside the exterior walls, such as balconies, terraces, or corridors are excluded.

Core Factor

The core factor, sometimes called the common area factor, is used to convert space quantity between usable and net rentable area. Normally that factor ranges between 10–15 percent. GSA defines common area factor in its standard solicitation as: "The Common Area Factor is a conversion factor(s) determined by the building owner and often applied by the owner to the usable area to determine the rentable square feet for the building."

Core factor is often confused with space efficiency. Core factor has nothing to do with space efficiency factors because core factor excludes the gross area of the building. It does not include the area taken by basements, penthouses, or exterior closure.

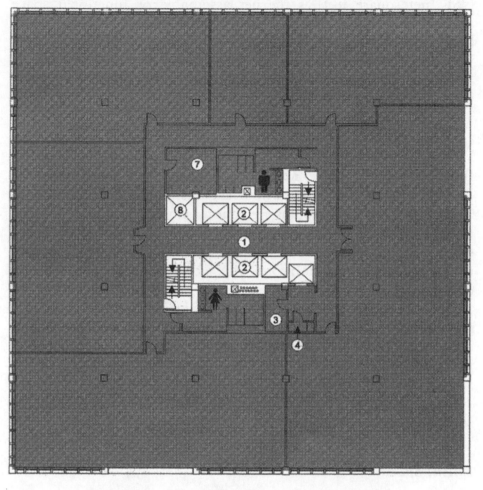

1. Lobby	2. Elevator	3. Electrical Room
4. Janitor	7. Fan Room	8. Ventilation Shaft

Figure 3.3 Illustration of floor rentable area for a typical upper-level floor

Converting Net to Gross

The cost engineer must convert net area program requirements to gross area building requirements. In doing so, the following formula is helpful to understand:

GSF = NSF + Support Space + Latent Space

NSF (net square feet) is the net assignable or functional space outlined as requirements or needs by the tenant. A potential problem to watch out for occurs in converting space from net to gross when the requirements specified by the tenant mistakenly include "support space," which is normally included in the efficiency ratio used to make the conversion.

Support Space

Building support space normally includes space to support the building population on each floor such as: public toilets, mechanical rooms, electrical rooms, telephone closets, janitorial closets, lockers for maintenance and custodial staff, building management storage, and loading docks. Support space also includes "circulation space" for stairs, elevator shafts, elevator lobbies, and fire exit corridors.

Private support space, for private tenant use in addition to that normally included in a building, should be included as net assignable space. The most common of this type of support space includes private toilets, telecommunications rooms, mail rooms, program storage space, mechanical space for computer rooms, security guard rooms, fitness space, and space for private circulation (normally a security-related requirement).

Latent Space

Latent space is the area of the building taken up by its exterior walls, fixed interior core partitions, and shafts. This type of space has always been included in the conversion factors and has never caused confusion.

The formula for converting net square feet to gross square feet is:

GSF = NSF/efficiency ratio

Figure 3.5b shows a sample conversion of NSF of space for a regional corporate headquarters building. The quantities of all three space types was computed from an actual bid designed in response to an SFO to show the breakdown between assignable, support, and latent space. This demonstrates that the efficiency ratios, when used in budgeting, account for all the space required to design the building.

Adjusting for Building Height

In budgeting any given project, the net assignable programmed space area is constant. The gross area to support that program, however, is not constant.

For example, assume that the NSF requirement is 250,000. Providing that area in a ten-story building will require more gross square feet than providing it in a one-story building. Space is saved in a one-story building because no space is required for stairs or elevators.

In addition, the size of each floor area influences the amount of gross square feet needed for the support areas.

Figure 3.4[2] provides a table of configuration factors that was developed for office buildings. It can be used to adjust the standard space efficiency ratios on most related space. The procedure to follow to determine which adjustment factor to use is as follows:

1. Determine the total NSF to be accommodated above grade. Divide that area by 0.70 to approximate the gross area above grade.
2. Determine the buildable site area in square feet. This area should represent the cost engineer's best judgment of the most likely above-grade building footprint considering all known site variables (setbacks, easements, site slopes, adjoining buildings, building heights, local code floor–area ratio, local architecture, etc.).
3. Divide the projected gross area by the buildable area and round to the next highest whole number. This represents the number of whole floors above grade.
4. Divide the approximate gross area by the number of whole floors to determine the approximate floor size.
5. Select the adjustment factor from Figure 3.4 using the approximate floor size and number of floors calculated as described above.
6. Use the selected factor to adjust the space efficiency ratios selected for each type of space.

Example:

Assumptions—

Office space efficiency factor = 0.75
NSF requirements = 203,000
Buildable area = 50,000

TYPICAL FLOOR SIZE IN SQUARE FEET

Sub-column order within each floor-size group: Custodial | Toilets | Mechanical | Horizontal Circ. | Vertical Circ. | Construction (for Percent of Total Support), and a single value for Configuration Factor.

BUILDING HEIGHT IN STORIES ABOVE GRADE		LESS THAN 12,000						12,000–18,000						18,000–25,000						25,000–35,000						OVER 35,000					
3–5 FLOORS	Percent of Total Support	7	6	20	36	7	24	7	7	21	33	8	24	7	6	21	35	8	23	7	6	22	35	8	22	7	5	23	35	8	22
3–5 FLOORS	Configuration Factor			0.92						1.00						1.02						1.00						1.01			
6–11 FLOORS	Percent of Total Support	7	6	22	33	8	24	7	7	23	31	9	23	7	6	23	34	8	22	7	6	24	33	8	22	7	5	25	33	9	21
6–11 FLOORS	Configuration Factor			0.93						1.01						1.03						1.01						1.02			
12–17 FLOORS	Percent of Total Support	7	6	23	32	10	22	7	7	24	30	11	21	7	6	24	33	10	20	7	6	25	32	10	20	7	5	26	32	11	19
12–17 FLOORS	Configuration Factor			0.94						1.02						1.05						1.02						1.03			
18–23 FLOORS	Percent of Total Support	7	6	23	30	11	23	7	7	24	27	12	23	7	6	25	30	11	21	7	6	26	29	11	21	7	5	27	29	12	20
18–23 FLOORS	Configuration Factor			0.91						0.98						1.01						0.98						1.00			
OVER 23 FLOORS	Percent of Total Support	7	6	24	29	12	22	7	7	25	25	14	22	7	6	25	28	13	21	7	6	27	27	13	20	7	5	28	28	13	19
OVER 23 FLOORS	Configuration Factor			0.90						0.97						1.00						0.97						0.98			

Figure 3.4 Efficiency ratio adjustment factors

Compute—

1. Approximate gross area

    GSF = 203,000/0.75 = 270,666 gsf

2. Approximate number of stories

    No. = 270,666/50,000 = 5.4 floors
    say 5 floors (plus 0.4 floor basement or penthouse)

3. Approximate floor size

    Size = 270,000/5 = 54,000 gsf each

4. Enter Figure 3.4, for a building height of 3–5 floors and a typical floor size of over 35,000 square feet
    Configuration factor = 1.01
    If the space efficiency factor to be used for a conference room in the building is selected as 0.66, then its efficiency factor is multiplied by
    1.01 from the table:

    0.66 × 1.01 = 0.67

Sample Space Program

Figure 3.5b is a sample space program for the office building headquarters. This format is available to download (see Appendix A). The file name is mspace.xls.

One only needs to fill in the shaded area of the format shown in Figures 3.5a and 3.5b. All cells but the shaded ones are protected. The steps to completing this format are:

1. Fill in the first three shaded cells (Figure 3.5a) that are at the top of the format file mspace.xls. This will compute the estimated number of stories just as described above. Then select the efficiency ratio for the predominate type of space from Figure 3.4 and enter it as shown in the shaded cell to the right.
2. Enter the program NSF.

Assumptions:			
Total Program Net Area	203,850	Eff. Ratio	
Assumed overall efficiency	0.75	Selected	1.01
Approximate gross area (sf):	271,800		
Buildable area (sf):	50,000		
Number of stories:	5		

Figure 3.5a Top part of mspace.xls

SPACE PROGRAM

Project: Office Building Headquarters
Location: Atlanta, Georgia

Areas	NSF	Efficiency Ratio	Ratio Adjustment	GSF Totals
Tenant Space				
Private Office	28,000	0.70	1.01	39,604
Open Office	65,000	0.75	1.01	85,809
Agency Suites	45,000	0.80	1.01	55,693
Conference	9,900	0.66	1.01	14,851
Library	1,700	0.76	1.01	2,215
Computer Room	14,000	0.61	1.01	22,724
Computer Lab/Testing	4,500	0.61	1.01	7,304
Storage Space	10,500	0.85	1.01	12,231
Training Space	7,300	0.70	1.01	10,325
Cafeteria	3,800	0.67	1.01	5,615
Auditorium	3,500	0.70	1.01	4,950
Coffee Bars	750	0.80	1.01	928
Fax/Printer/Copy Rooms	2,000	0.85	1.01	2,330
Loading Dock	2,200	0.90	1.01	2,420
Health Center	1,200	0.60	1.01	1,980
Security Control Center	500	0.80	1.01	619
Smoking Room	1,000	0.80	1.01	1,238
Vending/Break Room	3,000	0.80	1.01	3,713
Parking—interior		0.90	1.01	0
Other		0.80	1.01	0
Other		0.80	1.01	0
Totals	**203,850**			**274,549**
Support Space	**Percent**			
Horiz. Circulation	35.0			24,745
Vert. Circulation	8.0			5,656
Mechanical/Electrical	23.0			16,261
Toilets	5.0			3,535
Custodial	7.0			4,949
Construction	22.0			15,554
Total Support Space	**100.0**			**70,699**

Figure 3.5b Sample space program (mspace.xls)

3. Enter the efficiency ratios for each type of space from Figure 3.1.
4. Enter the percentages of support space from Figure 3.4.

The format sheet automatically calculates the GSF totals that are used to establish configuration described in the next chapter.

Notes

1. American National Standards Institute, *ANSI/BOMA Z65.1-1996*, Washington, D.C.
2. GSA Handbook, *Capitalized Income Approach to Project Budgeting*, Washington, D.C., U.S. General Services Administration.

4   Establishing Configuration

The Massing Diagram

When an architect is employed for design programming, the final product of this effort is the preparation of a massing diagram for the facility (see Figure 4.1). The massing diagram combines all the elements of the owner's goals, functional relationship study, code and zoning requirements, and site analysis in general building envelopes.

This massing diagram was prepared for a project bid by the author for development of a new office headquarters building in Atlanta, Georgia. All further examples in subsequent chapters of this book will be based on this sample project for illustrative purposes.

Figure 4.1 Massing diagram

In the absence of an architect, the author provides a statistical configuration technique in this chapter that produces all the architectural system quantities necessary for parameter budgeting without getting into the aesthetic issues surrounding the massing diagram. It is good to have a massing diagram but not absolutely necessary for budgeting.

Configuration Decisions

Determining configuration involves making three decisions or assumptions regarding the number of floors, floor perimeter, and floor heights.

Number of Floors

Number of floors is a strong factor in determining cost. Historical data is often provided in three cost ranges: low rise (one to four stories), mid rise (five to ten stories) and high rise (over ten stories).

Often the SFO or tenant program will dictate the size of the ground floor by indicating who needs that type of space, that is, commercial or walk-in tenants. Parking and support space needs for mechanical and electrical may dictate below-grade areas.

The community and/or the desire to present a good image may dictate the requirement for a "clean roof" appearance, thus placing equipment within the building that would otherwise be on the roof, or requiring a penthouse.

Each type of space has its layout efficiencies. For example, the optimum size for office space layout is 25,000 to 35,000 gross square feet on a floor.

Shape

Building shape also strongly influences cost. Figure 4.2 shows three shapes for a constant floor area. The single shape variable is the length of building perimeter.

- Shape A, the rectangle, is what might be expected of a commercial quality building.
- Shape B, with an interior court, provides the most expensive shape as it has the largest perimeter.

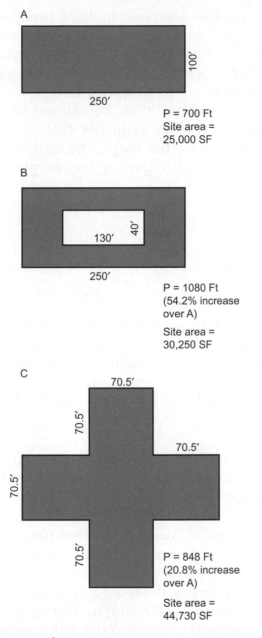

Figure 4.2 Shape vs. perimeter

- Shape C is midway between A and B. It is called an "articulated" perimeter. It has a perimeter 15–20 percent greater than shape A, giving the architect much more freedom of design expression.

Therefore, if one does not have a building configuration suggested by an architect, one can budget to permit some architectural freedom simply by choosing to budget for an increased perimeter. Of course, the perimeter selected will affect the amount of heating and air conditioning provided as discussed later.

Also, it must be observed that the amount of site area needed for shape C is approximately double that of shape A. This is logical because the wider the shape, the further out the building setback lines are located, requiring facility placement on larger sites.

Height

Height of a facility influences the type of mechanical systems that can be employed.

- Floor-to-floor heights of 10–11 feet would require an all-electric building or high-velocity flat duct systems.
- Distances of 11–12 feet would allow use of hydronic systems as well as above.
- Distances of 12–14 feet allow low-velocity VAV duct systems as well as above.

The type of program space influences building height. Space types such as large lobbies, courtrooms, auditoriums, and cafeterias often require high ceiling heights that may protrude through one or more floors.

Many local zoning ordinances restrict building height. Therefore, in order to fit a given program within the height limit, the floor size must be enlarged on the site. One should find out about these requirements at the budget stage. This type of data can be quite important; for example, in the case of speculative developed buildings, this could restrict the amount of rentable area that can be provided on the site and make the project uneconomical.

Site Assumptions

Before massing or configuration decisions are made one must perform site area analysis.

Site Area Analysis

Site analysis falls into three categories; either:

1. the developer or owner has a site;
2. several potential sites exist; or
3. no site has been sought.

Therefore, site area analysis either documents conditions of an existing site or the same data sets the goals for the characteristics of a site to acquire. Things to describe for each site include:

- size;
- demolition requirements;
- contour—hilly, sloped, flat;
- vegetation—wooded, grassy, desert;
- utilities—gas, water, electric, sewer;
- drainage—water table expectations;
- soil—rock, clay, gravel, sand;
- foundation—types of surrounding buildings;
- traffic—access and egress;
- environmental concerns.

Where the site is given and the local zoning restrictions are known, one has less latitude in influencing the budget. However, when the site is not predetermined, there is wide budget latitude.

The site area and related site budget as well as the construction budget are interdependent. In this case, the cost engineer must develop a hypothetical building configuration. However, before proceeding with configuring the hypothetical building, one should perform the following type of analysis.

Code and Zoning Analysis

Perform a quick code and zoning analysis. Pertinent enforcement authorities having jurisdiction over the site area should be contacted to obtain copies of code and zoning requirements. A listing of all requirements affecting the project should be prepared, such as:

- FAR (floor area ratio);
- parking;

- building height;
- setbacks;
- landscaping percentages;
- type of construction allowed;
- zoned use permitted.

Community Requirements

Review community requirements for factors influencing the configuration decision. These include:

- height of neighboring buildings;
- water table elevation;
- location of utilities;
- terrain features;
- future expansion requirements;
- traffic circulation, street access;
- master plan (if available).

The knowledge developed from this review will aid in the selection of an appropriate building footprint and subsequent building height and basement depth determination.

If not available from a design programming effort, the cost engineer should develop a rough site sketch noting the above factors and outlining the building footprint used for budget development. This would then serve as a reference document in future site selection or budget modification.

Determining Site Area

The size of a site suitable to fit a specific tenant program is a function of the following elements, which, when summed, provide the minimum acreage necessary:

- building footprint area (a function of the number of stories planned to achieve the programmed square footage);
- exterior parking (a function of the number of cars needed to meet zoning requirements for the facility or to meet stated tenant requirements);
- exterior landscaping area (a function of zoning requirements);

- setback distances (a function of zoning and/or security requirements);
- other site amenities—plaza, child care playground, etc. (a function of tenant requirements).

Configuration Objectives

Configuration must be determined for each individual structure in the project if more than one structure is involved.

The objectives in determining configuration are for use in determining design parameters and for computing cost for various budget elements. Ultimately, the cost engineer will obtain from this data the number of floors, the volume of the facility, and the surface areas of the roof and exterior wall.

Statistical Configuration Process

Figure 4.3a illustrates a systematic format originally developed by GSA,[1] and slightly modified by the author, to guide the computation of statistics of the building to be budgeted.

Figure 4.3b is a sample building configuration for the office building headquarters. This format is provided with this text (see Appendix A). The file name is mconfig.xls. One only needs to fill in the shaded area of the format shown in Figure 4.3b. All cells but the shaded ones are protected. For convenience in discussing how this format

Building Configuration Computation Sheet

Select a building perimeter code from the list below and enter in following box: [4]

Code	Desired Shape
1	1:1 Rectangular
2	2:1 Rectangular
3	3:1 Rectangular
4	Articulated
5	1:1 open central court
6	2:1 open central court
7	3:3 open central court

Figure 4.3a Perimeter selection

Project: Office Building Headquarters, Atlanta, GA

Building Configuration

ITEM	1ST FLOOR	UPPER FLOORS	BASEMENT	PENTHOUSE	TOTAL BUILDING	SITE AREA
2-1　GROSS AREAS (sf)	51,500	206,049	9,000	8,000	274,549	940,460
2-2　BUILDABLE AREA (sf)	50,000	50,000	50,000	8,000		
2-3　NO. OF FLOORS	1	4	1	2	6	
2-4　NO PENTHOUSE					1	
2-5　TYPICAL FLOOR AREA (sf)	51,500	51,512	9,000	8,000		
2-6　SLAB ON GRADE AREA (sf)					51,500	51,500
2-7　SITE IMPROVEMENT AREA (sf)						888,960
2-8　TOTAL FLOOR CONST (sf)					223,049	
2-9　FIRST FLOOR AREA (sf)			51,500			
2-10　BASEMENT EXT ROOF (sf)			0			
2-11　F-F HEIGHT (ft)	18.00	13.00	12.00	12.00		
2-12　BLDG. EXT. HEIGHT (ft)	18.0	52.0	12.0	12.0	82.0	
2-13　PERIMETER (ft)	1,270	1,270	402	378		
2-14　CLOSURE AREA (sf)	22,860	66,040	4,824	4,536		
2-15　EXT. CLOSURE (sf)					93,436	
2-16　VOLUME (cf)	927,000	2,678,637	108,000	96,000	3,809,637	
2-17　TOTAL NET AREA (sf)					203,850	
2-18　BLDG SUPPORT AREA (sf)					70,699	
NOTES						

Figure 4.3b Configuration statistics

works, the author has assigned imaginary numbers to each column. These are:

- first floor—(Column 1);
- upper floors—(Column 2);
- basement—(Column 3);
- penthouse—(Column 4);
- total building—(Column 5);
- site area—(Column 6).

The preceding data box is provided at the top of the computer format (Figure 4.3a). Select the desired shape of the building by selecting its code and entering the code number in the shaded box. For our sample project, code 4 was selected. This selection will automatically provide the perimeter distance in lineal feet for an *articulated* perimeter building. The number 1,270 was calculated and entered by the computer on line 2-13 of the format. The program uses the floor area (line 2-5) and the shape to calculate the perimeter.

Computations in the format are not rounded out beyond whole numbers unless otherwise shown. Referring to the computer format file for Figure 4.3b, instructions to calculate configuration data follow:

1. Line 2-1. Enter the total site area in Column 6. (All areas must be entered in square feet.)
2. Line 2-1. Using the space program prepared in Chapter 3, allocate the gross area of the building between first floor, upper floors, basement, and penthouse (Columns 1–4). The total building area (Column 5) will be computed automatically and should match the total GSF computed for the space program.

    • In the example, the author saw that mechanical/electrical space was 16,261 square feet in the program, rounded it off, and allocated 9,000 square feet of it to basement space and 8,000 square feet to penthouse.
    • Next, from the previous computation to determine the ratio adjustment factor used in the space program to modify the efficiency factors, the author let the first floor be 51,500 square feet with the balance of the area being in the upper floors.

3. Line 2-2. Enter the buildable area. It is that ground area portion of the site permitted by code to contain the building.

    • The buildable area (footprint) for the building above grade should be equivalent to the area used in Chapter 3 for adjusting building height.
    • The buildable area for the basement, Column 3, may exceed the floor area depending on the judgment of the cost engineer. For example, on a large site, underground parking often extends outside the building line when going deeper is not economical or feasible.

Normally, a maximum of three basement levels is assumed and the footprint area must be large enough to accommodate requirements within three levels. If this is not possible, basement area requirements must be reduced. A separate above grade parking structure might then be considered.

- • The buildable area for the penthouse, Column 4, is independent of the building above grade footprint. However, it cannot be larger than the roof area unless it is a multi-story penthouse.

4. Line 2-11. Select and enter the floor-to-floor heights based on style and experience. See Chapter 7 for guidance. Normally, the first floor is higher than the upper floors. The author has selected 18 feet for the first floor and a 13-foot structural distance for the upper floors in the example.
5. Line 2-17. Enter the program total net square foot area in Column 5.

All the remaining computations on the building configuration format are performed by the computer.

Computer Computations

Here is what the computer is doing:

Line 2-3. The gross areas (Line 2-1) are divided by the buildable areas (Line 2-2) and round to the next highest whole number for entering on Line 2-3. These values represent the number of levels required for each basic area type. If the number of basement levels exceeds three, return to Line 2-2 and consider this carefully by reviewing site soil and ground water conditions.

Line 2-5. The respective gross areas (Line 2-1) are divided by the number of levels at Line 2-3 to determine the average floor size for each basic component area.

Line 2-6. The larger area at either Columns 1 or 3 (Line 2-5) is taken as the total slab on grade area.

Line 2-7. Site improvement area is calculated as the difference between the site preparation area and the above grade footprint area (subtract Line 2-6, Column 5, from Line 2-1, Column 6).

Line 2-8. Floor construction area or the area of supported floors is calculated as the difference between the total gross area of construction and the slab on grade area (subtract Line 2-6, Column 5, from Line 2-1, Column 5).

Line 2-9. A basement roof is required for any area of basement outside the line of the building above grade footprint. A special traffic topping (often a waterproof plaza deck) is assumed to be required for this area at grade. Therefore the first floor area from Line 2-5, Column 1, is entered here.

Line 2-10. In Column 3, subtract Line 2-9 from Line 2-5 and enter the result if greater than zero. If greater than zero, this quantity is the special traffic topping area.

Next, are the calculations for height, closure, and volume:

Line 2-12. This line multiplies the respective floor-to-floor heights at Line 2-11 times the number of levels for each component at Line 2-3 and enters the project heights (in feet) at Line 2-12. Check the local code to ensure that the building height just calculated does not exceed it.

Line 2-13. Building perimeter is entered by the computer. See below for more detailed explanation of this procedure. One can unprotect the cells on this line and override the computer input if desired.

Line 2-14. Closure area is determined by multiplying the respective floor heights, Line 2-12, by the perimeters, Line 2-13, to obtain the closure areas for each component (basement, penthouse, building).

Line 2-15. Exterior closure area is calculated as the sum of the closure areas for the building above grade plus the penthouse (Line 2-14, Columns 1, 2, and 4). The remaining closure area (Line 2-14, Column 3) represents basement wall area.

Line 2-16. Building volume (in cubic feet) is obtained by multiplying component heights at Line 2-12 times their respective average floor areas at Line 2-5. Columns 1–4 are summed and entered in Column 5 to provide total building volume.

Line 2-18. Building support area is determined by subtracting Line 2-17 from Line 2-1. Check to see that this matches the space program support area.

Building Perimeter

The length of building perimeter for any given area is controlled by the shape of the building. The most efficient (shortest) perimeter for any given area is the circumference of a circle. On occasion, if a circle shape is desired then one can solve for the circumference knowing the ground floor area.

However, the most common shape of building is the rectangle. When both sides are equal, the rectangle is known as a square and is expressed as having a 1:1 shape ratio. Buildings that are twice as long as they are wide have a 2:1 shape ratio. The longest rectangle one should use in budgeting is a shape ratio of 3:1 because buildings longer than that are less structurally rigid and need extraordinary cost to make rigid against local wind and seismic loads.

Buildings with open central court areas are often used for buildings with large gross floor areas. This shape provides for the maximum amount of wall area, outside the building and inside the court. This shape is often used to provide a large amount of floor space on a window wall for light and/or ventilation. Sometimes the courtyard is covered over into an atrium. This only affects HVAC loads. The inside courtyard wall must still be counted as perimeter in budgeting.

Somewhere between the minimum perimeters of rectangular shapes and the maximum perimeters of rectangular shapes with courtyards fall other shapes like triangles, semi-circles and zigs and zags. Rather than specifying these shapes and pretending to be an architect, the author budgets for an *articulated* perimeter. The

Building Footprint Area (FPA)	Rectangular			Articulated	Open Central Court		
	1:1	2:1	3:1		1:1	2:1	3:1
3,000 sf	220	235	250	310	330	350	375
6,000 sf	310	335	360	435	465	500	540
9,000 sf	380	410	440	530	570	615	660
10,000 sf	400	440	480	560	600	660	720
20,000 sf	565	605	650	790	850	910	975
30,000 sf	690	745	800	970	1035	1120	1200
40,000 sf	800	860	920	1120	1200	1290	1380
60,000 sf	980	1055	1130	1370	1470	1580	1695
80,000 sf	1130	1225	1320	1580	1695	1840	1980
100,000 sf	1265	1360	1460	1770	1900	2040	2190
120,000 sf	1385	1500	1610	1940	2080	2250	2415
140,000 sf	1500	1620	1740	2100	2250	2430	2610

Figure 4.4 Building perimeters (lineal feet)

articulated perimeter allows for freedom of shape for indention and projection of a building perimeter without specifically defining it.

As explained, Line 2-13 of the computer format will compute and display the proper perimeter lengths for various length to width ratios for various shaped buildings. One can override the cells on Line 2-13 by unprotecting them and entering their own perimeter length. Figure 4.4 provides a table computed for each of the shapes discussed with the amount of building perimeter to budget for each.

Note

1. GSA Handbook, *Capitalized Income Approach to Project Budgeting*, Washington, D.C.: U.S. General Services Administration.

5 Architectural and Structural Quantities

Introduction

Most of the architectural and structural quantities necessary for estimating are now available from the building configuration model developed in Chapter 4, Figure 4.3b.

Review of the author's database with this text reveals the units of measure available for pricing each major item. Other units of measure can be used for various items of work, but these then must be priced in accordance with those units.

(01) Foundation

To estimate several foundation quantities requires making design assumptions and calculations as well as a quick look ahead to substructure and superstructure systems.

- Foundation area (FPA) can be obtained from Figure 4.3b, Line 2-5, Column 1.
- Foundation perimeter (LF) for grade beams or foundation drain system can be obtained from Figure 4.3b, Line 2-13, Columns 1 and 3.

Should the cost engineer desire to budget foundations based upon the "working load" in KIPS at column locations, the following additional data is available:

- Minimum live loads are determined from code. For budgeting most buildings use Figure 5.1.[1]
- Dead loads should be based on structural and wall system selections as documented (see R.S. Means[2] for dead load material weights).

- Wind load is determined from geographical location (see Chapter 2). Wind resistance area can be based on one-half of closure area; Figure 4.3b, Line 2-14, Columns 1, 2, and 4.
- Seismic zone can be determined from Figure 5.2 below as extracted from the National Building Code.[3]

Reference Aids		R15.1–100	Live Loads		

Table 15.1–101 Minimum Design Live Loads in Pounds per S.F. for Various Building Codes

Occupancy	Description	Minimum Live Loads, Pounds per S.F.		
		BOCA	ANSI	UBC
Armories		150	150	
Assembly	Fixed seats	60	60	50
	Movable seats	100	100	100
	Platforms or stage floors	100	100	125
Commercial & industrial	Light manufacturing	125	125	75
	Heavy manufacturing	250	250	125
	Light storage	125	125	75
	Heavy storage	250	250	100
	Stores, retail first floor	100	100	75
	Stores, retail upper floors	75	75	
	Stores, wholesale	125	125	100
Court rooms		100		
Dance halls	Ballrooms	100	100	
Dining rooms	Restaurants	100	100	
Fire escapes	Other than below	100	100	
	Multi or single family residential	40		
Garages	Passenger cars only	50	50	50
Gymnasiums	Main floors and balconies	100	100	
Hospitals	Operating rooms, laboratories	60	60	
	Private room	40	40	
	Wards	40	40	
	Corridors, above first floor	80	80	
Libraries	Reading rooms	60	60	
	Stack rooms	150	150	125
	Corridors, above first floor		80	
Marquees		75	75	
Office Buildings	Offices	50	50	50
	Lobbies	100	100	
	Corridors, above first floor	80	·	100
Residential	Multi family private apartments	40	40	40
	Multi family, public rooms	100	100	
	Multi family, corridors	80	·	100
	Dwellings, first floor	40	40	40
	Dwellings, second floor & habitable attics	30	30	
	Dwellings, uninhabitable attics	20	20	
	Hotels, guest rooms	40	40	40
	Hotels, public rooms	100	100	
	Hotels, corridors serving public rooms	100	100	
	Hotels, corridors	80	80	
Roofs	Flat	12–20		20
	Pitched	12–16		
Schools	Classrooms	40	40	40
	Corridors	80	80	100
Sidewalks	Driveways, etc. subject to trucking	250	250	250
Stairs	Exits	100		100
Theaters	Aisles, corridors and lobbies	100	100	
	Orchestra floors	60	100	
	Balconies	60	100	
	Stage floors	150	150	
Yards	Terrace, pedestrian	100	100	

BOCA = Building Officials & Code Administration International, Inc. National Building Code
ANSI = Standard A58.1
UBC = Uniform Building Code, International Conference of Building Officials
* Corridor loading equal to occupancy loading.

Figure 5.1 Table of live loads

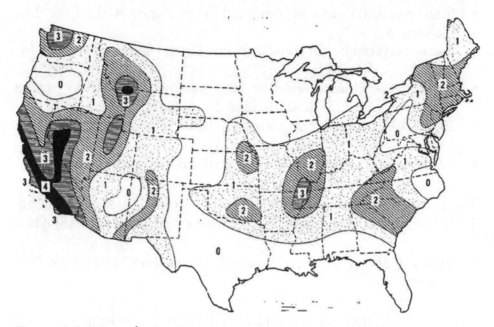

Figure 5.2 Map of seismic zones, contiguous 48 states

Foundation assumptions should be based on the specific site when known, soil tests if available, or knowledge regarding the types used on adjacent or nearby structures. Otherwise, one can budget based on the following:

- Provide grade beams in seismic zones to tie standard foundations.
- Provide strip footings in lieu of grade beams under basement walls if soil will support 3–10 KSF.
- Soil capacity limits story height on standard foundations:

  - 3 KSF, up to five stories;
  - 6 KSF, up to ten stories;
  - 10 KSF, up to 15 stories.

- Figure piles, caissons, or mat foundation for everything else.

(02) Substructure

The substructure of a building includes the three major elements of slab on grade, and basement excavation and basement walls (if any).

- Slab area (FPA) can be obtained from Figure 4.3b, Line 2-6, Column 5.
- Excavation volume (CY) can be obtained from Figure 4.3b, Line 2-16, Column 3, divided by 27.
- Basement wall area, sheeting, and shoring area (SFSA) can be obtained from Figure 4.3b, Line 2-14, Column 3.
- Depth of basement for stairwell (LFR) can be obtained from Figure 4.3b, Line 2-12, Column 3.
- Waterproofing areas for slab and walls are known from the above.

The substructure seismic zone is known from foundations above. To be computed is the substructure supported floor area (SFA):

- Figure 4.3b, Line 2-3, Column 3, minus one, times Line 2-5, Column 3.

To be computed are the number of stairs for exiting from the substructure to grade. The formula to be used is:

N = GSF of largest floor / (273 × W)

where:

N = number of stairwells
W = width of stairs in inches
Minimum W = 36" Standard W = 44" Largest W = 60".

Normally, for economy, the width of stairwell for substructure floors will match the width for the upper floors assuming the stairwell is continuous. From the above formula, using the standard 44-inch width, one stair is adequate for a small basement area up to 12,000 square feet.

The final item to be assumed for pricing purposes is the bay spacing for the substructure. This is a qualitative issue. Guidance on selecting this is a matter of building style as is discussed in Chapter 7. The range of bay sizes provided for in the author's database varies from a 25 × 25 foot bay to a maximum of 35 × 40 foot bay. The types of floor structure—steel, composite, joist slab, and waffle slabs—are provided in the database with prices varying depending on live load, span, and number of supported floors. Many other unique combinations of materials and spans can be designed at additional cost. The guidance provided by the author's database are normal and will perform.

When providing substructure depth, one should be guided by program requirements as well as adjacent or nearby structures. In addition:

- Slab thickness is a function of use and loading:

  - 100 psf, foot traffic—4";
  - 200 psf, storage—5–6";
  - 500 psf, traffic—6–8";
  - 1500 psf, machinery—8–12".

- Sheeting and shoring depend on excavation layback area around the site:

  - sand, gravel—1:1 layback;
  - clay—1:2 layback;
  - unstable—shoring required.

- Basement walls –12" masonry is best assumed good for only one story below grade.

(03) Superstructure

Building superstructure includes the major elements of floor construction, roof construction, canopies, and stair construction (including handrails).

To be computed is the superstructure supported upper floor area (UFA):

- Figure 4.3b, Line 2-8, Column 4 provides the total supported floor construction. If there is more than one basement level, subtract the area of supported floor computed above for the substructure.

Other data also known are:

- Structural roof area can be obtained from Figure 4.3b, Line 2-5, Column 2.

- Structurally supported basement roof area can be obtained from Figure 4.3b, Line 2-10, Column 3 (if any).
- Height of building for stairwell (LFR) can be obtained from Figure 4.3b, Line 2-12, Columns 1 and 2. This is necessary because the author's database prices stair per flight (FLT) based on a ten-foot height plus add-on expense for greater heights per flight.

Known from foundation data above, are the superstructure bay size, and seismic zone. Number of stairs can be computed using the same formula as provided for substructure stairs. The super-structure type should be based upon the prevalent type used in the area or found economical to the region.

- Wood frame and load bearing structures are normally limited to no more than four to six stories depending on fire code and loading.
- If structural steel is selected, it must be assumed to require fireproofing.
- Live loads should meet the applicable building code.
- Bay size designation is important. Smaller sizes reduce structural cost but also influence layout flexibility:

  - 25 × 25—poor;
  - 30 × 30—average;
  - 35 × 35—good;
  - 40 × 40—excellent.

Bay size should normally be the same in both substructure and super-structure. Sometimes bay sizes in the substructure are smaller than in the superstructure in order to accommodate parking. When this occurs, one must add approximately 10 percent to the substructure cost to provide for load transfer beams for wider spans on the upper floors.

In addition, sometimes the tenant's program will specify bay sizing because it impacts their ability to make an efficient space layout. Often bay spacing is influenced when the requirement is expressed as a desire to have a five-foot planning module.

(04) Exterior Closure

The exterior closure consists of the exterior walls, windows, and exterior doors of all types.

- The total closure area of the building (XCA) can be obtained from Figure 4.3b, Line 2-15, Column 5.

The materials, style, and amount of fenestration is another qualitative issue for a building (see Chapter 7). The author's database prices an adequate range of choices to document scope, such as: fixed sash, casement, horizontal pivoted, vertical pivoted, double hung, and curtain wall window styles with varying frame materials and glazing quality.

However, to meet energy conservation criteria walls must limit heat transmission to 200 btu/hr/lf maximum. Therefore, the maximum fenestration allowed is:

15–20%—single glazed;
35–40%—double glazed.

- The amount of door and window area (XDA) included in the total closure area (XCA) is a function of the percent of fenestration (%fen) allowed. Then:

XDA = XCA × %fen.

- The amount of solid wall area (XWA) on the building exterior is:

XWA = XCA − XDA.

- Closure area of penthouse (XWA) can be obtained from Figure 4.3b, Line 2-14, Column 4.

The number of fire exit doors is a code requirement. They always must be at least equal to the maximum number of stairs (from the substructure or superstructure).

Exterior closure budgeting of materials and glazing should be guided by the surrounding environment and the desire to budget capital for future energy savings.

- For energy conservation design, the "U" factor of the exterior wall must be in the range of 0.06–0.10.
- Energy conservation design standards for lobby areas require vestibule entrances and/or revolving doors.

- Design for handicapped accessibility standards requires automatic entrance doors.

(05) Roofing

The roofing category includes the full system of roof covering, insulation, flashings and trim, skylights and hatches.

   The author's database provides for various quality of roof coverings: 3-ply, 4-ply, membrane, EPDM, metal, and fiberglass. Other materials can be added to the database by the reader. Roof area in square feet for roofing material and insulation area can be obtained from Figure 4.3b, Line 2–5, Column 2. Sometimes this is measured in squares (SQ) where 100 SF = 1 SQ.

- The penthouse roof area need not be added because its footprint area is *already* contained within the typical upper floor area used above.
- Roof perimeter (LF) for parapet walls and flashing can be obtained from Figure 4.3b, Line 2–13, Columns 2 and 4. The author normally doubles these quantities to provide for flashing and trim at the perimeter cap and for counter-flashing at the base.

The roofing "U" factor must be in the range of 0.05–0.10 to be considered energy conserving. The author's database provides a variety of materials and insulating quality from which to select, such as: fiberboard, fiberglass, foam board, lightweight concrete, and perlite fill.

- If significant skylight area is included in the budget, its total area must be included in the fenestration percentages set forth above to meet energy conservation standards.
- If a large amount of roof mounted equipment is contemplated, then maintenance walkways should be considered in the budget.

(06) Interior Construction

Interior construction consists of partitions, finishes, and specialties.

Partitioning

The author's database provides for three types of fixed partitions: drywall, plaster, and concrete masonry unit (CMU). Each of these are associated with a partitioning ratio and a general fire rating. The database assumes a nine-foot ceiling height for non-rated partitions and a 13-foot slab-to-slab height for rated partitions.

A partitioning ratio is the degree of density of partitioning in a given space. Partitioning ratios are expressed as lineal feet of partitioning to square feet of space (lf:sf). Thus a 1:12 ratio would provide one lineal foot of partition for every 12 net square feet of space. The common ratios provided in the database are:

- core, support spaces—1:5;
- individual spaces—1:12;
- open landscaped space—1:100.

The reader can customize the database by developing other partitioning ratios and adding them as desired.

- When budgeting for partitioning, the sum of the area used should be the full gross area of the building. That is, put some partitioning in the budget, at the proper density, for all the program space plus the support space.
- Use the net floor areas from the space program (Chapter 3) and group the areas by partition ratio and type.
- The following support spaces should almost always be figured as masonry because of the need for fire resistant construction and/or soundproofing:

  ○ mechanical/electrical rooms;
  ○ shafts/toilets;
  ○ elevators;
  ○ stairwells;
  ○ walls dividing fire zones.

- Corridors, and walls for other spaces to be permanently located during design, can be figured as gypsum board walls, with corridor walls fire rated.

- Interior doors are included within the partition pricing based upon the ratio of 1:X nsf (one door for X net square feet of space partitioned). Common ratios are:

  ○ core, support spaces—1:100;
  ○ individual spaces—1:150;
  ○ open space—1:4000.

The database contains three special types of partitions: demountable, folding, and toilet partitions. Review the types of program space to determine if any of these should be budgeted.

- Normally, open office space can be specified as demountable. And, where flexibility of reorganization is a high project goal, most of the individual spaces that do not involve special plumbing work are often budgeted with demountable partitioning.
- Look to the types of program space for folding partitions. Often conference rooms, training rooms, auditoriums, hearing rooms, cafeteria, and dining space are divided with folding partitions where functional flexibility is a project goal.
- To determine the number of toilet partitions, look to the fixture unit count of plumbing fixtures calculated in the next chapter.

Finishes

The author's database provides for the selection of finishes either by type of space or by type of material.

The easiest budgeting method is to select finishes by type of space using net square feet (NSF). This will be explained first.

NSF procedure

Interior construction quality and finishes depend largely upon the type of functional space being provided. The database contains prices for three quality levels of space: commercial quality, high quality, and superior quality.

The quality level selected affects the grade of materials for door hardware, floor covering, ceiling finish, and wall finishes normal and appropriate to the type of space. One can mix these quality levels to suit the program being budgeted.

- The area of program space (NSF) to be finished is as provided for in Chapter 3.
- The total space should match the budget for partitioning discussed above.
- All support space (NSF) to be finished is as developed in Chapter 3.

A percentage of the circulation space is, for example, the front lobby and elevator lobbies. A distinction in quality level can then be allocated to a portion of the circulation space.

Material area procedure

If one were to budget by type of material one would have to quantify and select different materials for walls, floors, and ceilings. This could certainly be accomplished by making a finish schedule in the form of a grid, listing all the program and supporting space types and areas down one column, and then calculating separate grids of quantities (or percentages) by material types for floors, ceilings, and walls.

The areas of program and support space to be used are as explained above. Then, the guidance in calculating quantities would be:

- Percentages in each category: i.e. partitions, doors, floors, ceilings, and walls should add up to 100 percent.
- Ceiling and floor finishes generally relate directly to the type of space programmed.
- The total finished area (TFA) budgeted (floors, ceilings, and walls) should equal the amount determined from the following equation:

$$TFA = 2 \times GSF + 2 \times PSF + XWA.$$

- Partition square feet (PSF) would need to be determined in order to allocate the proper finishes. This is computed based on the partition ratios depending upon the degree of density of partitioned space desired.

Specialties

The third element of the interior construction budget component of UNIFORMAT is to include specialties.

These are architectural specialties as listed below. Do not confuse them with fixed and movable equipment and furnishings discussed in the next chapter under UNIFORMAT element 11.

- For lockers and toilet accessories, look to the program and plumbing requirements to judge these quantities. The author's database estimates toilet accessories based upon SETS. One set is figured for each lavatory in the plumbing budget.
- Special handrails can be an expensive item. This can be quantified based upon the number of stairs previously calculated above for budgeting and their height. Do not forget to double the height to include railing on both sides of the stairwell.
- Look to the type and quantity of program space for miscellaneous counters, cabinets, and closets.
- Estimate chalk and tack boards and storage shelving based on the program space requirements.

Notes

1. R.S. Means, *Assemblies Cost Data*, 1998, Norwell, MA: Reed Construction LLC.
2. Ibid.
3. *The BOCA National Building Code*, 1987, Country Club Hills, IL: Building Officials and Code Administrators International Inc.

6 Mechanical and Electrical Quantities

Introduction

This chapter explains the computation of budget estimating quantities for UNIFORMAT elements 07 through 12.

(07) Conveying System

Conveying systems in the author's database include elevators for passenger, freight, and hospital use as well as moving stairs and walks, escalators, and conveyors.

Passenger Elevators

The normal capacity of passenger elevators can be budgeted at 3,500 pounds. The type of elevator generally is determined by the number of floor levels served:

- Traction geared elevators are suggested from four to 13 floors at speeds ranging 200–500 fpm.
- Gearless traction elevators are used in high-rise buildings at speeds from 600–1,000 fpm.
- Hydraulic elevators avoid the need for penthouses but have a maximum rise of 70 feet at 100 fpm. Avoid budgeting them above four stories because of the need for increased service and maintenance.

The number of elevators is determined by building population, waiting time, speed criteria, and function of space.

- In general, the elevator ratio 1:X people (one elevator for X number of people) would be:

  high level of service—1:500;
  fair level of service—1:750.

- When determining the number of elevators for a certain number of occupants, visitors should also be included.
- Separate, dedicated elevators, regardless of population, may be needed for restaurant, courtroom, theater, or similar spaces located above grade level.
- A more precise way to determine the number of passenger elevators is through the formula:

N = PfT / 300E

where:

  N = Number of elevators
  P = Design Population
  f = Peak factor, f = 16.5% of building population when all occupants start work at same time, f = 14.5% when they start work at staggered hours
  T = Elevator round trip time (seconds) on morning peak
  E = Normal number of persons per car on morning peak
    E = (.80 × C) / 150
  C = Car capacity in pounds.

- The length of rise (LF) for each elevator can be obtained from Figure 4.3b, Line 2-12, Columns 1–4.

The number of landing openings (LO), the parameter for estimating purposes, is calculated by adding the number of floors (stops) for each elevator.

Do not forget to add an additional stop for each basement level if one of the passenger elevators serves the basement. If the basement contains a parking garage, then a separate bank of passenger elevators for the garage should be added to the budget.

- Normally, garage elevators are 2,500 pound hydraulic type.
- The number of passengers is equal to the parking capacity of one level.

Freight Elevators

A good rule of thumb is to provide one freight elevator for every three passenger elevators, or for every 75,000 NSF of space.

- To maintain chillers on the roof, freight elevator capacity should be 8,000 pounds. Otherwise, budget for 4,000 pounds.
- Freight elevators should be budgeted to serve every level, from basement to penthouse.

Escalators

Escalators are used where there is a need to carry 600 or more people between floors. Their carrying capacity is 5,000–8,000 people per hour.

(081) Plumbing

Plumbing system quantities depend upon configuration data, the number of occupants and visitors, and the type of space programmed.

The author's database establishes two price ranges for plumbing depending on the relative expected location of the fixtures on a typical floor. The price ranges are called *central core* and *spread fixtures*. If most toilet rooms are stacked above each other vertically from floor to floor, price them as *central core* plumbing. If, however, the space program dictates rooms with special plumbing that will potentially be scattered all over the floor, away from the central core, then price those fixtures as *spread fixtures*. The number of fixtures is a function of the following factors:

- Standard number of fixtures is based on permanent occupancy, distributed by floors. If occupancy is not known, one can assume the following type of occupancies:

  - 100 NSF/person—based on fire code;
  - 135 NSF/person—based on GSA standards;
  - 160 NSF/person—based on plumbing code.

- The ratio of men to women is normally 50/50 unless specified otherwise by the program (often the case in a GSA SFO).
- Check user program requirements for the number of permanent staff and visitors as explained in Chapter 2.

NUMBER OF MEN*/WOMEN	WATER	LAVATORIES
1 – 15	1	1
16 – 35	2	2
36 – 55	3	3
56 – 60	4	3
61 – 80	4	4
81 – 90	5	4
91 – 110	5	5
111 – 125	6	5
126 – 150	6	**
>150	***	

*   In men's facilities, urinals may be substituted for ½ of the water closets specified.
**   Add one lavatory for each 45 additional employees over 125.
*** Add one water closet for each 40 additional employees over 150.

Figure 6.1 Fixture unit criteria—office space

Figure 6.1[1] provides the number of fixture units (FU) for office type space based on building population per floor.

- Use net square feet of space (NSF) per floor to compute occupancy.

Figure 6.2 provides the number of fixture units for other types of special spaces. The number of special fixtures is based on the number of visitors and the needs of special types of functional space in the program.

- Number of FUs for special purpose plumbing is based on the space program, Chapter 3.
- Include requirements for support space as defined in Chapter 3.
- Number of floor drains for mechanical space and interior parking space are calculated as one-half of an FU.

Roof Drains

The area of roof to be drained is obtained from Figure 4.3b, Line 2-5, Column 2.

- The maximum rainfall in inches is obtained from geographical data as explained in Chapter 2.
- Figure 6.3 provides the roof drain coverage to compute the number of drains.

OCCUPANCY	MEN	WOMEN
Executive office	2/office	2/office
Employee assembly	3:1–15 4:16–35 7:36–55 >55 people, 3/40	2:1–15 3:16–35 6:36–55 >55 people, 2/40
Public assembly	3:1–100 5:101–200 8:201–400 >400 people, 4/500	3:1–100 4:101–200 6:201–400 >400 people, 3/300
Restaurant	3:1–50 4:51–150 9:151–300 >300 people, 3/200	2:1–50 3:51–150 7:151–300 >300 people, 3/400
Dormitory	3:1–10 4:11–12 5:13–20 6:21–24 >25 people, 3/25	2:1–8 3:9–12 4:13–16 5:17–24 >25 people, 2/15
Training	3/40	2/30
Prisons	2/cell	2/cell
Exercise room	2/room	3/room

Note: written as (FU:range of people)
or (FU/per number of people)

Figure 6.2 Fixture unit criteria—special space types

Rainfall	Area of Roof Covered per Drain	
Inches	4" Drain	5" Drain
1	18,400	34,600
2	9,200	17,300
3	6,100	11,500
4	4,600	8,600
5	3,700	6,900
10	1,800	3,400

Figure 6.3 Roof drain coverage

To use Figure 6.3, select the rainfall and roof drain size desired to determine the coverage ratio. Divide the area of the roof by the coverage ratio to determine the number of roof drains. Divide the number of roof drains by one-half to represent them as fixture units (FU) for pricing.

Plumbing Model

Figure 6.4 is a sample plumbing model program for the office building headquarters. This format is provided with this text (see Appendix A).

Project: Office Building Headquarters				
Location: Atlanta, Georgia				
Criteria			0.40	0.60	Total
135 sf/person		NSF	No. men	No. women	People
Basement		9,000	27	40	67
Floor 1		38,970	115	173	289
Floor 2		38,970	115	173	289
Floor 3		38,970	115	173	289
Floor 4		38,970	115	173	289
Floor 5		38,970	115	173	289
			0	0	0
		203,850	604	906	1,510

Quantities	WCs	Lavs	Urinals	Total Fixtures
Men's toilets				
Basement	1	2	1	
Floor 1	4	5	2	
Floor 2	4	5	2	
Floor 3	4	5	2	
Floor 4	4	5	2	
Floor 5	4	5	2	
Women's toilets				
Basement	3	3		
Floor 1	7	7		
Floor 2	7	7		
Floor 3	7	7		
Floor 4	7	7		
Floor 5	7	7		
Totals	59	65	11	135

Special Requirements	Water	Sinks	Drains	WC/Lav	
Private toilets	–	–	–	10	10
Cafeteria	4	4	2		10
Auditorium	2				2
Coffee Bars	5	5	5	–	15
Loading Dock			1	–	1
Health Center	5	2		4	11
Vending/Break Room			5	–	5
Computer Room	4		12	–	16
Shower/Locker Room			4	8	12
Other					0
Other					0
Building Requirements					
Drinking fountains	–	–	–	–	0
Janitor closets		5	5	–	10
Mechanical rooms			2		2
Hose bibs	8	–	–	–	8
Roof	–	–	20	–	20
					257

Figure 6.4 Sample plumbing model

The file name is mplumb.xls. Filling in the shaded area of the format with the proper selection of data will provide this model. The steps to completing this format are:

1. Fill in the criteria in the third row (the number of sf/person occupancy).
2. Fill in the ratio of men to women in the third row (decimal form as shown, which totals to 1.00).
3. List the number of occupied floors and their respective NSF areas.

At this point, the computer calculates the number of people (men and women individually) per floor. Use this data to enter Figure 6.1 and select the number of water closets and lavatories for men and for women.

4. Enter these on the format as shown.
5. Reduce the number of men's toilets by one-third and make them urinals (whole numbers of fixtures only and a minimum number of one each per room per floor).

Review the special program requirements.

6. Use Figure 6.2 to assist in selecting the number and types of fixtures to enter in the format.

Determine the general building requirements.

7. Enter these on the format.

(082) HVAC

HVAC system quantities depend upon the area of the building envelope, the occupancy, and the percentage of fenestration. From geographical location one can determine the winter and summer outside design temperature. From owner criteria, the inside design temperature can be specified. By selecting the quality level of exterior wall and roof one can determine the insulating or "U" values desired.

From this data one can calculate the block load BTUH of heating required and the TONS of air conditioning required. By reviewing the program one can add additional air conditioning for

HVAC COMPUTATIONS **Project:** Office Headquarters Building

Location:	Atlanta, Georgia			
Summer DB T2 =	92		Interior T1 =	76
Winter DB T3 =	22		Interior T4 =	72
Humidity Ratio W2 =	0.0160		Humidity W1 =	0.0102

COOLING LOAD Degree Days = 1,589

Envelope	R	FACTOR UNIT		QUANT MEAS	T2–T1	BTUH
Roof	30.0	0.033	"U"	51,512 SF	16	27,473
Exterior Wall	19.0	0.053	"U"	62,230 SF	16	52,404
Exterior Glass	1.2	0.840	"U"	26,670 SF	16	358,588
Other	–	–	"U"	SF	16	–

Solar					FACTOR	
Additional Roof Gain		0.033	"U"	51,512 SF	55	94,439
N. Glass Exposure				6,668 SF	37	246,698
S. Glass Exposure				6,668 SF	111	740,093
E. Glass Exposure				6,668 SF	219	1,460,183
W. Glass Exposure				6,668 SF	219	1,460,183
Shading Credit				3,907,155 BTUH	–0.43	(1,680,077)

Interior						
Ventilation Air (cfm/person)	20					
latent		1.1	BTU/CFM	72,780 CFM	16	1,280,928
sensible		0.0058	W2–W1	72,780 CFM	4840	2,043,080
People		550	BTUH/PER	3,639 PER		2,001,450
Lighting & power		3	WATT/SF	244,995 SF	3.41	734,985
Computer room		10	WATT/SF	14,000 SF	3.41	140,000
Other		–	WATT/SF	SF	3.41	–
					TOTAL BTUH =	8,960,426
					TONS =	747

Equivalent Electricity = KWH/YR = 5,010,481

HEATING LOAD Degree Days = 3,095

Envelope	FACTOR UNIT		QUANT	MEAS	T4–T3	BTUH
Roof	0.033	"U"	51,512	SF	50	85,853
Exterior Wall	0.053	"U"	62,230	SF	50	163,763
Exterior Glass	0.840	"U"	26,670	SF	50	1,120,588
Other	–	"U"	–	SF	50	–

	RATE					
Infiltration	1.1	BTU/LF	26,670	LF	50	1,466,850
Ventilation	1.1	BTU/CFM	72,780	CFM	50	4,002,900
					TOTAL BTUH =	6,839,955
					MBH =	6,840

| Equivalent Electricity = | | | | | KWH/YR = | 3,464,992 |
| Equivalent Gas = | | | | | CCF/YR = | 140,662 |

Figure 6.5 Sample HVAC model

special purpose loads such as computer rooms, auditoriums, exhaust fume hoods, cafeterias, and court space.

In addition, the CFM of ventilation for enclosed parking can be calculated from program and configuration data.

Block Load Computations

The author has provided the format (see Appendix A) to assist in making HVAC block load computations. The file name is mhvac.xls. One only needs to fill in the shaded cells as shown in Figure 6.5 (HVAC Model) for the sample headquarters office building project to compute the TONS of air conditioning and MBH of heating. In addition, the format provides an estimate of the annual energy consumption to operate the HVAC equipment.

Figures 6.6a[2] and 6.6b provide a listing of the HVAC data necessary at the budget stage for use in completing the HVAC model format to calculate loads. Included are data for the basic roof types, glazing types, shading, solar gain, occupancy activities, and humidity ratios.

The data shown in Figure 6.6 is presented in the order in which it is selected and used to complete the HVAC model format. Following this data is an explanation of the procedures to complete the HVAC format.

A. HUMIDITY RATIOS	
East coast	0.016 – 0.017
Inland	0.013 – 0.015
Desert, high altitude	0.006 – 0.008
West coast	0.010 – 0.012

B. GLAZING TYPES	"R" factor
Single glass	0.96
Solar or Reflective glass	1.2
Insulating glass – double	1.64
Insulating glass – triple	2.27

C. ROOF TYPES	Cooling Load Factor
(without suspended ceiling below)	
Insulated metal	80
Built-up roofing system	70
Roof terrace	40
(with suspended ceiling below)	
Insulated metal	70
Built-up roofing system	55
Roof terrace	30

Figure 6.6a HVAC computation factors

D. SOLAR GAIN FOR GLASS					
	Btu/sf by Latitude				
	20	24	32	36	40
North	48	45	37	36	35
East	212	213	219	218	216
South	43	46	111	131	149
West	212	213	219	218	216

E. SHADING COEFFICIENT	No Shading	Venetian Blinds	Roller Shades	Drapes
Clear glass	0.95	0.64	0.39	0.6
Tinted, heat absorbing glass	0.69	0.57	0.36	0.46
Reflective coated glass	0.40	0.33	NA	0.46
Insulated double glass	0.88	0.57	0.37	0.52
Insulated coated glass	0.30	0.27	NA	0.34

F. OCCUPANCY ACTIVITIES	Btu/hr/person
Auditorium, theater	350
Office, hotel	480
Cafeteria, restaurant	520
Commercial, retail	800
Light factory	1,020
Bowling, dancing	1,300
Heavy factory	1,600
Gymnasium	2,000

G. VENTILATION REQUIREMENTS	Minimum cfm per person
Office space	20
Conference rooms	25
Lobbies, waiting areas	10
Auditoriums	5
Gymnasiums	20
Library	7
Laboratories	15
Toilets	15
Locker rooms	30
Cafeterias	10
Corridors	15

Figure 6.6b HVAC computation factors

HVAC Computation Procedures

The steps to completing the mhvac.xls format as illustrated by Figure 6.5 are:

1. Enter the summer and winter design temperatures (in degrees Fahrenheit) for the geographical location as explained in Chapter 2. Normally these are based on ASHRAE's 97.5 percent value.

2. Enter the interior summer and winter design temperatures.

   - These are dependent on user desires, not location.
   - For energy efficient design meeting GSA standards these are:

     ○ 76 degrees, summer;
     ○ 72 degrees, winter.

3. Enter the proper humidity ratio from Figure 6.6a, Item A.
4. Enter the cooling load degree days (only necessary if you want energy consumption).
5. Enter the envelope "R" factors.

   - Pick the glass value from Figure 6.6a.
   - Pick insulating glass to qualify for energy efficient design.

6. Enter the envelope quantities.

   - Pick roof and wall quantities computed in Chapter 5.
   - Enter glazing quantity computed in Chapter 5.
   - For energy efficient design, glazing quantity with its corresponding "R" factor must meet criteria explained in Chapter 5.

7. Enter the glazing "R" factor from Figure 6.6a, Item B, based on the type of glazing computed in Chapter 5.
8. Enter the glass exposure areas for North, South, East, and West.

   - The area of glass by exposure should be in the same ratio as the perimeter ratio determined in Chapter 4.
   - To be conservative, unless you know the orientation of the building, match the largest glass area with the largest solar factor.

9. Enter the additional roof gain solar factor from Figure 6.6a, Item C, based on the type of roof computed in Chapter 5.
10. Enter the solar gain for glass for each compass direction from Figure 6.6b, Item D.
11. Enter the shading credit. This credit is a negative value determined by subtracting 1.0 from the desired shading coefficient from Figure 6.6b, Item E. Of course, the type of glazing and window equipment used to determine the shading coefficient should match that included elsewhere in the budget.
12. Enter the proper people load factor from Figure 6.6b, Item F.

13. Enter the number of people in the quantity column.

   • The number of people should include visitors.

14. Enter the main ventilation rate for the building from Figure 6.6b, Item G.
15. Enter the lighting and power watts per square feet.

   • For office space this can range from 2.5 to 3.5 watts/sf.

16. Enter the NSF of space for lighting and power.
17. Enter any special power requirements in watts/sf (such as for the computer room in the sample project).
18. Enter the NSF for special power areas.
19. Enter the heating load degree days (if energy consumption is desired).

The computer calculates the total tonnage for air conditioning and MBH of boiler heating requirement without spare or standby capacity. The 747 TONS computed in the example is the peak load. Additional tonnage might be added to this number if spare or standby equipment is desired; 25 percent spare is a reasonable amount, however, some owners want as much as 50 to 100 percent spare capacity.

The cost of an HVAC system is dependent on several issues listed below in increasing order of quality. The author's database provides for the following:

• Type

   ○ rooftop package units;
   ○ heat pump system;
   ○ air-handling unit (AHU) system;
   ○ (VAV) variable air volume system.

• Control

   ○ 1 zone per room;   ○ 1 zone per 3,500 GSF;
   ○ 1 zone per 1,500 GSF;   ○ 1 zone per 5,000 GSF.

(083) Fire Protection

Fire protection requirements are normally set by building code or by the owner's insurance company. The GSA and some other organizations have a policy to provide sprinkler protection regardless of code.

The author's database provides for wet pipe, dry pipe, preaction, and deluge type sprinkler systems at three hazard levels: light, ordinary, and extra hazard. The most normal uses of these systems are:

- wet pipe—heated interior space;
- dry pipe—space where freezing can occur such as loading docks and parking garages;
- preaction—space where alarm is desired to protect equipment before water flows such as computer rooms;
- deluge—space having significant quantities of combustible fuel.

In addition, buildings with sprinkler systems normally have standpipe systems in the stairwell, and buildings can have standpipes with hose reels for firefighting without having sprinklers. Quantities for estimating these systems can be obtained as follows:

- Gross area for sprinkler calculations can be obtained from Figure 4.3b, Line 2-1, Column 5.
- Height for each standpipe riser can be obtained from Figure 4.3b, Line 2-12.
- One standpipe per stairwell is provided when standpipes are used.
- Number of floors or stations for each standpipe riser can be obtained from Figure 4.3b, Line 2-3. Multiply this by the number of stairs.

(084) Special Mechanical

Review Appendix B, UNIFORMAT item 084, to see if any special systems should be included in the estimate. Special systems are normally a function of the type of space defined in the program as well as special requirements of the owner.

Special mechanical systems must be specifically listed and quantified. Review the types of space programmed to see if they have any special mechanical requirements such as loading dock air curtains, garage exhaust, laboratory space, kitchen exhaust, etc.

(09) Electrical System

The first step is to estimate the total primary incoming load for electrical service distribution. The author's database uses the total AMPS of incoming service to price the main switchboard, substation, panels, transformers, and service buses.

Figure 6.7 illustrates the electrical block load computation for the sample project. This format is provided with this text (see Appendix A). The file name is melec.xls. All but the shaded cells are protected. Fill the shaded cells as follows:

1. Enter the program and support area net square footage in the second column.
2. Enter the percentage distribution of support area (same as determined for Figure 3.5b).
3. Enter the TONS of HVAC computed (Figure 6.5).
4. Enter the horsepower of any fire pumps.

Electrical Block Load Computation

Project: Office Headquarters Building
Location: Atlanta, Georgia

PROGRAM AREA		NSF	LIGHTING WATTS/SF	POWER WATTS/SF	CONNECTED KW	DEMAND FACTOR	DEMAND KW
Net Usable Area							
Private Office		28,000	1.5	0.5	56.0	75%	42.0
Open Office		65,000	1.5	1.0	162.5	75%	121.9
Agency Suites		45,000	1.5	0.5	90.0	75%	67.5
Conference		9,900	1.5	0.5	19.8	75%	14.9
Library		1,700	1.5	0.5	3.4	75%	2.6
Computer Room		14,000	2.0	8.0	140.0	100%	140.0
Computer Lab/Testing		4,500	2.0	1.0	13.5	75%	10.1
Storage Space		10,500	1.0	0.5	15.8	75%	11.8
Training Space		7,300	1.5	0.5	14.6	75%	11.0
Cafeteria		3,800	2.0	4.0	22.8	75%	17.1
Auditorium		3,500	1.0	0.5	5.3	175%	9.2
Coffee Bars		750	1.5	0.5	1.5	75%	1.1
Fax/Printer/Copy Rooms		2,000	1.5	1.0	5.0	75%	3.8
Loading Dock		2,200	1.5	0.5	4.4	75%	3.3
Health Center		1,200	1.5	0.5	2.4	75%	1.8
Security Control Center		500	1.5	0.5	1.0	75%	0.8
Smoking Room		1,000	1.0	0.5	1.5	75%	1.1
Vending/Break Room		3,000	1.0	0.5	4.5	75%	3.4
Parking—interior			0.2	0.2	0.0	50%	0.0
Other					0.0		0.0
Other					0.0		0.0
Totals		203,850			563.9		463.2
Support Area	Percent	70,699					
Horiz. Circulation	35.0	24,745	1.0	0.5	3.2	50%	1.6
Vert. Circulation	8.0	5,656	1.0	0.5	37.1	50%	18.6
Mechanical/Electrical	23.0	16,261	0.5	1.0	8.5	50%	4.2
Toilets	5.0	3,535	1.0	0.5	24.4	50%	12.2
Custodial	7.0	4,949	0.5	0.5	3.5	50%	1.8
Construction	22.0	15,554					
HVAC	747 Tons				1,494.0	100%	1,494.0
HVAC VAV boxes		242,734		0.5	121.4	100%	121.4
Elevators		242,734		1.0	242.7	100%	242.7
Fire Pumps	35 Hp				38.5	50%	19.3
Spare Capacity		274,549		1.0	274.5	50%	137.3
Parking, roads		399,750		0.2	80.0	50%	40.0
Totals					2,327.8		2,093.0
			SUBTOTAL KW =		2,891.7		2,556.1
	Spare Capacity	25 %			722.9		639.0
			TOTAL KW =		3,614.7		3,195.2
For 3 phase supply voltage =	480		Factor = 0.83	AMPS =	4,347.8		3,849.6
			SERVICE AMPS SAY =	4,500.0			

Figure 6.7 Electrical model—sample project

5. Enter the gross square foot area for HVAC VAV boxes, elevators, and spare capacity.
6. Enter the site paved area for lighting exterior parking:

   • number of cars times 325 square feet per car.

7. Enter the percent of spare incoming capacity:

   • use 25 percent if not specified by user.

8. Enter the incoming service voltage:

   • check with local utility;
   • normally 480 volts for large buildings;
   • could be 277 volts for smaller buildings.

The format provides the block electrical load based upon energy efficient design standards for each type of space. This standard has been provided on a watt/sf basis in protected cells of the format. The reader can unlock the format and input higher electrical loads if energy efficient design is not desired.

The computer format calculates the total connected kilowatts (KW) and converts KW to AMPS based upon the desired three-phase voltage. The KW to AMPS conversion formula used by the format is:

AMPS = (KW × 1,000) / (Volts × 1.732).

Other useful formulas are:

Motors – 1 HP = 746 WATTS
Rule of thumb – 1 KVA = 1 HP.

For information, the energy conservation design standards of GSA for office space limits the installation of lighting and power to 7.0 watts/GSF rather than the previous normal range of 10.5 watts/GSF as shown below:

Lighting—	2.0	vs.	2.0
Power—	1.0	vs.	2.5
HVAC—	2.0	vs.	4.0
Elevators—	1.0	vs.	1.0
Spare—	1.0	vs.	1.0
Total	7.0	vs.	10.5

Lighting and Power

The author's database provides pricing for lighting and power on a square foot basis, include lighting and power for both program space and support space.

Prepare the lighting estimate by grouping space requirements on a square foot basis as follows:

- first, by type of lighting: fluorescent, incandescent, mercury vapor, high pressure sodium (HPS) or high-intensity discharge (HID);
- next, by footcandle (FC) intensity or watts per square foot.

Prepare the tenant power estimate by grouping space requirements on a square foot basis as follows:

- first, by type of system: duplex, under floor duct, poke thru, power pole, conduit system, or under-carpet power system;
- next, by watts per square foot.

Add power for the building for HVAC, elevator, and spare capacity based on gross square feet (GSF) area.

Special Electrical

Review Appendix B, UNIFORMAT item 093, to see if any special electrical systems should be included in the estimate. Special systems are often a function of the type of space. Review the types of space programmed to see if they have any special electrical requirements, such as: auditorium sound system, office computer network, etc.

Also, some systems are quality requirements such as:

- fire alarm systems, number of STA based on a minimum of one per stairwell floor landing;
- building grounding system estimated in LF of copper conductors and NUMBER of ground rods;
- emergency power system estimated in KW or KVA—emergency power needs for fire safety should represent no more than 10 percent of the total demand;

- emergency power for operation of critical space and equipment should be identified and added to the fire safety needs;
- floor raceway system estimated in NSF;
- telephone system estimated in STA, including PABX equipment.

(11) Equipment

Review Appendix B, UNIFORMAT item 11, for equipment that might be included for the building in general. Review the type of special space for equipment and furnishings normally necessary for space such as for courts, classrooms, computer, auditoriums, laboratories, cafeterias, etc.

- Equipment and furnishing needs must be specifically listed and quantified. Review the author's database to ascertain pricing parameters.
- Review the program space to ascertain what normal built-in equipment might be provided to outfit the space. Most projects have equipment such as:
  - window washing;
  - waste handling;
  - loading dock;
  - parking equipment;
  - postal equipment;
  - telephone equipment.

- Most buildings also have:
  - floor directories;
  - signage;
  - blinds or shades.

After equipment is listed, the special mechanical and electrical connection costs (084 and 083), if any, should be included under those categories.

(121) Site Preparation

The site preparation area (SF) is determined from Figure 4.3b, Line 2-1, Column 6. To convert to ACRES, divide by 43,560 square feet per acre.

(122) Site Improvements

The total site improvement area (SF) is determined from Figure 4.3b, Line 2-7, Column 6. This area represents the full area of the site less the footprint area of the building. This area consists of requirements for both hard paving and landscaping as well as the area under which utility services are routed in easement areas from off-site.

Compute the site area required for hard paving including:

- access roads;
- parking spaces;
- sidewalk area;
- plaza;
- terrace areas;
- hard paving.

- Exterior parking requirements can be computed at 325 square feet per car. One-way traffic lanes can be figured at 12 feet wide, two-way at 24 feet wide.
- The minimum plaza area (SF) around the building is provided by Figure 4.3b, Line 2-10. Additional plaza area may be determined by program needs. Unless indicated otherwise, the author budgets a small entrance plaza of 1,000 square feet.
- Sidewalks, three feet wide, can be based upon the perimeter of the building plus access distance from parking lots and/or main road.

Landscaped area:

- Subtract the total hard paved area from the site improvement area to determine the soft area remaining for landscaping.

Other site improvements:

- Fencing can be based on the perimeter of the site and/or perimeter of parking or structures to be protected. Where shape is not known, use a 1:1 ratio based on the formula:

$$P = 4 \times \text{(square root of fenced area)}.$$

- Review Appendix B, UNIFORMAT item 122, for a listing of special site improvement elements. Review the design program for desired amenities such as ball fields, track, picnic areas, lighting, etc.

(123) Site Utilities

Attempt to judge the average distance (LF) of utility runs from the source to the building line. Estimate all distances making right angles rather than assume diagonal routing of piping.

- Figure 6.8[3] provides guidance in determining the size of sewer and storm systems based upon plumbing fixture unit and rainfall data.

SEWER PIPE SIZE

Pipe Diameter (inches)	Max. No. Connected Fixture Units
4	180
5	390
6	700
8	1,600
10	2,900
12	4,600
15	8,300

STORM SEWER PIPE SIZE

Pipe Diameter (inches)	Paved and Roof Area Surfaces Maximum Rainfall (inches/hour)		
	2	4	6
5	6,680	3,340	2,230
6	10,700	5,350	3,570
8	23,400	11,500	7,600
10	41,400	20,700	13,800
12	66,600	3,300	22.200
15	109,000	59,500	39,700

Figure 6.8 Sewer pipe size

- For fire protection systems, sprinklers and/or outside hydrants, figure a minimum pipe size of six inches, looped around the perimeter of each structure at least 50 feet away. Add to that quantity the distance to the site boundary.
- Unless otherwise known, budget all electrical using underground duct bank, four-way minimum system. This provides for one power feeder and one spare, plus one telephone feeder and one spare.

In some locations, the budget solution might consist of wells and pump houses for water; package treatment plants or septic systems for sewage; and oil storage for diesel generators.

Notes

1. General Services Administration, standard SFO paragraph 6.5 (Solicitation for Offers), Washington, D.C.
2. Source—Abstracted from ASHRAE, *Cooling and Heating Load Calculation Manual*, 1984, Atlanta, GA: American Society of Heating, Refrigeration, and Air Conditioning Engineers, Inc.
3. Abstracted from *ASPE Data Book*, vol. 1, *Fundamentals of Plumbing Design*, 1983–1984, Rosemont, IL: American Society of Plumbing Engineers.

7 Construction Cost Estimating

Introduction

Chapter 2 discussed the key cost driver project information necessary for budgeting. Chapter 3 discussed the methodology to compute space quantities. Chapter 4 discussed the methodology to calculate configuration-related quantities useful for systems estimating. Chapters 5 and 6 provided detailed guidance to determine each relative system parameter quantity following the UNIFORMAT system of estimating. At this point in following the text, one has prepared and assembled sufficient project documentation to permit the preparation of a fairly detailed estimate.

This chapter provides additional discussion of the concepts behind development of parameter quantities, the use of building quality guidelines, the use of the estimate as a scope document, the methodology of UNIFORMAT estimating and the author's database that can be used to compute project construction cost.

The need for parameters

In determining quantities, the traditional method of measuring is parameter cost. Probably the most misunderstood and misused term is *parameter*. The dictionary defines a parameter as an arbitrary constant, each of whose values characterizes a member of a system.

The most common way to estimate a new building is by the cost per square foot. This classical parameter is really not a parameter at all. Cost is neither constant, nor does it vary in a consistently predictable pattern, nor does it characterize any particular system.

The major problem with using costs per square foot as a parameter is that the cost for that unit of measure is constant only for one class or type of building at a particular time. Retrieval and reapplication of cost per gross square foot ($/GSF) data requires use of extreme care, good judgment, and complete understanding of the separation of classes inherent between differing cost per gross square foot statistics.

The user of cost per square foot data must know more about the basis for the data in order to separate its applications between the buildings inherent in the statistics. For example, knowing the cost per gross square foot for constructing a residence does not help one to price a ten-story office building. Parameters at the building level are difficult to develop, qualify, quantify, and store for future use.

Similarly, cost per gross square foot pricing for systems such as exterior closure, plumbing, mechanical, and electrical systems is not very helpful. However, parameters at the systems level are easier to develop in a meaningful way than is cost per gross square foot. Generally, parameter units of measure can be developed based upon some term or characteristic of the system to be priced. Table 7.1 provides some common system-level parameters used for building construction.

System Parameters

With proper system parameters there should be less fluctuation between unit costs than there is between comparable square foot costs. Also, assembling several different parameters to form an estimate will provide a greater degree of budget accuracy than placing sole reliance on one cost per gross square foot.

Several system parameters also rely on the use of basic design criteria to quantify. Budgeting using criteria is preferred because it provides another cost control mechanism.

Quality and Quality Standards

The major external influence on building quality is the site location and its surrounding environment. If one is planning to build in or around old town Alexandria, Virginia, for example, the quality and style of building permitted in the area probably will have to be compatible with the brick facades and New England style prevalent in the area. If not budgeted on that basis, potential cost problems could occur should the design be forced into emulating historic preservation in order to obtain a building permit. The major internal influence on building quality is the desire and pocketbook of the

Table 7.1 Units of measurement

System	Measure	Meaning
01 Foundations		
011 Standard Foundations	FPA	Footprint area (square feet)
02 Substructure		
021 Slab on Grade	SFSA	Square feet of surface area
022 Basement Excavation	CY	Cubic yards
023 Basement Walls	SFSA	
03 Superstructure		
031 Floor Construction	SSA	Supported surface area (square feet)
032 Roof Construction	SSA	
033 Stair Construction	FLT	Flight
04 Exterior Closure		
041 Exterior Walls	SFSA	
042 Exterior Doors and Windows	SFSA	(or EA—each)
05 Roofing	SQ	Square (100 square feet)
06 Interior Construction		
061 Partitions	PSF	Partition square feet
062 Interior Finishes	SFSA	Square feet of surface area
	NSF	Net square feet
07 Conveying Systems	LO	Landing openings
08 Mechanical		
081 Plumbing	FU	Fixture unit
082 HVAC	TONS	One ton = 12,000 btuh
	MBH	1,000 btuh (heating system measure)
083 Fire Protection	HEAD	Number of sprinkler heads
	STA	Stations (for standpipe systems)
09 Electrical		
091 Service and Distribution	AMP	Amperes of connected load
092 Power and Lighting	GSF	Gross square feet
10 General Conditions and Profit	PCT	Percent
11 Equipment		
121 Site Preparation	ACRE	Acre
122 Site Improvement	SY	Square yard
123 Site Utilities	LF	Lineal feet

owner, user, or developer. The goals and reasons for the project play a major role in the selection of quality.

1. If the project is for a corporate owner, then issues such as corporate image, space efficiency, low operating and maintenance costs, and employee amenities play a larger role.
2. If the project is for long-term rental income then issues such as marketing image, security, and layout flexibility play a larger role. Operating cost is not as much of a quality concern if escalation of utility costs is passed through to the tenant.
3. If the project is for speculative development, what counts is the inclusion of quality features that will gain tenants quickly and will last until the time the owner plans to sell.

Defining Quality

Building quality is an elusive element, often considered to be in the eye of the beholder. The greatest misconception is the belief that higher cost always relates to higher quality. Of course this notion is false. In many instances, there is no fixed, direct relationship between cost and quality. One can have a high-quality town house and a low-quality mansion.

It is true that copper flashing costs more than galvanized steel flashing. However, both in many cases have acceptable quality. The difference in these products is their performance. Quality is directly related to performance. If a product works exactly as designed and expected, it is considered a high-quality product. The same is true for building systems. If the elevator system performs as designed in terms of response time, speed, mean time between failure, and carrying capacity, then its quality is deemed high.

Quality products do break down and do require maintenance because a certain type of performance is expected. Owners probably couldn't afford a product that would never require maintenance. Therefore, the amount of maintenance and repair expected for a system is part of the definition of its quality.

Quality Standards

As part of the project scope defining process, one must document the building quality and system qualities being budgeted. To assist in providing a quality baseline for budgeting, GSA developed and published the building quality guidelines shown in the three pages of Figure 7.1.[1] These standards and criteria should used as the basis for budgeting projects in the absence of differing information from the specific location.

Even though four building styles are provided—monumental, federal, corporate owner, and commercial investor—one could mix

Quality Level	1	2	3	4
Style	Monumental	Federal	Corporate Owner	Commercial Investor
Economic life	100	50	25	20
Space efficiency	.60	.70	.75	.80
Lobby space	Concourse Courtyard Gallery Atrium	Concourse Courtyard Atrium	½ concourse ¼ courtyard	
Parking style	Self	Self	Self & stack	Stack
Special space	Cafeteria Shops Courts Auditorium ADP Vaults Public use	Cafeteria Shops Courts Auditorium ADP Vaults Public use	Cafeteria Shops Auditorium ADP	Snack bar Shops
Art in Architecture	Yes	Yes	Sometimes	No
Setbacks (urban site)	Yes	Yes	Sometimes	No
Percent site used (urban site)	50	50	75	75–90

NOTICE

The primary purpose of this document is to show significant differences which may occur between different owners. Any owner may include any of these elements based on geographical location, local codes, and benefits.

Figure 7.1 Quality standards

Quality Level	1	2	3	4
01 Foundations				
Site conditions	Poor bearing Often wet	Poor bearing Often wet	Average	Excellent Normally dry
03 Superstructure				
Floor height—1st	15–25	15–20	15	12
other	14–15	13	12	11
Spans	35–40	30–35	25–30	20–25
Module	5 foot	5 foot	No	No
UBC importance factor	1.5	1.25	1.0	1.0
Live load	100 psf	100 psf	80 psf	50 psf
Stair width	5'0"	4'0"	3'6"	3'0"
04 Exterior closure				
Material quality	Granite/marble	Limestone/brick	Brick/precast	Curtain wall
Energy efficiency	Yes	Yes	Possible	No
Fenestration	Bronze	Anodized aluminum	Aluminum	Aluminum or steel
Glazing	Solar insulated	Insulated	Tinted	Clear—single
Automatic doors	Yes	Yes	Sometimes	No
Operable windows	Pivoted	Pivoted	Sliding	Hopper
% fenestration	Up to 50	Up to 40	Up to 30	Up to 25
Revolving doors	Yes	Yes	Sometimes	No
05 Roofing				
Material quality	Slate/tile/ copper	4-ply/metal	Single membrane	Single membrane
Skylights	Yes	Yes	Possible	No
Energy efficient	Yes	Yes	Yes	No
Aesthetic screening	100%	75%	75%	25%
06 Interior Construction				
Material quality Lobbies	Marble	Stone	Masonry	Masonry
	Terrazzo	Ceramics	Paneling	Paneling
Corridors		Precast	Carpeting	VCT
Partitioning	Masonry Plastered	Demountable Sound-resistant	Vinyl clad Permanent	None
Ceilings	Plastered Acoustic	ICB	Splined	Lay-in
LF partitions/SF	1:20	1:20	1:30	1:40
Time rated ceilings	No	No	Yes	Yes
07 Conveying Systems				
Handicapped	Yes	Yes	Sometimes	No
Escalators	Yes	Possible	Possible	No

Figure 7.1 (continued)

08 Mechanical

HVAC

Dual fuel	Yes	Yes	No	No
Split loads	Yes	Yes	Sometimes	No
Expansion capability	Yes	Yes	Sometimes	No
5-foot module	Yes	Yes	No	No
Control	Individual	Individual	Small zones	Zone by exposure

Plumbing

Handicapped	Yes	Yes	Sometimes	No
Wall mtd fixtures	Yes	Yes	Sometimes	No
Sprinklers	Yes	Yes	Sometimes	No

09 Electrical

Expansion capability	Yes	Yes	Sometimes	No
Emergency generator	Yes	Yes	Sometimes	No
Motor control center	Yes	Yes	Sometimes	No
Security system	Yes	Yes	Sometimes	No
Lighting panel breakers	Bolt on	Bolt on	Plug in	Plug in
5-foot module	Yes	Yes	No	No
Fire alarm system	Yes	Yes	No	No

11 Equipment

Outfit special purpose space	Yes	Yes	Yes	No
Mail chutes	Yes	Yes	Yes	Yes
Package chutes	Yes	Yes	No	No
Window washer	Yes	Yes	Yes	No
Loading dock leveller	Yes	Yes	Sometimes	No

Sitework

Landscaping	Yes	Yes	Little	Very little
Plazas	Yes	Yes	Little	No
Material	Slate	Slate	Brick	Concrete
Flagpoles	Yes	Yes	Sometimes	No
Handicapped ramps	Yes	Yes	No	No

Figure 7.1 (continued)

the selection of criteria between the styles in budgeting a project. Not all criteria have to come from one column.

Estimating Level of Detail

The recommended level of detail for a construction budget is to use system parameters with system unit costs. For items of scope unfamiliar to the estimator, then add these using the component cost method.

Systems Cost

The summation of systems level cost estimates creates the gross cost estimate. The dictionary defines a system as a set or arrangement of things so related or connected with other subsystems as to form a unity or whole.

Within this context, one must be careful in properly using historical cost data at the systems level of detail. For example, using the above definition, cost for an air conditioning system should include the cost for the controls and wiring as well as for the chiller equipment and distribution equipment. However, in some estimating data one will find that controls and wiring fall under electrical cost categories rather than under the mechanical cost account. The UNIFORMAT standards deal with this issue and many other similar issues.

Component Cost

Within system level estimates are components. A component is a part, constituent, or ingredient of the whole. For example, lighting fixtures are a component of a lighting system, and a lighting system is a component of an electrical system.

Identification and quantity calculation of the major components in a system will generally lead to the creation of a cost parameter for the system.

The Framework for Cost Accounting

Regardless of the specific procedures used to develop the budget estimate, or the level of detail supporting it, a standard framework is beneficial because it:

- ensures consistency in the preparation of budgets over time and from project to project;
- allows cost information related to different stages of project development to be compared uniformly;
- provides a checklist for the budgeting process to reduce error and improve accuracy;
- reduces ambiguity and facilitates clear communication between all members of the project team;
- provides a continuing frame of reference to develop and maintain effective project cost control.

A standard framework therefore provides a unifying effect through each stage of project development, from budget inception through construction and ultimately into occupancy.

The latest state-of-the-art in accounting structure for budget and project estimating is the 12-division structure called "UNIFORMAT." The UNIFORMAT classification system was developed through a joint effort of GSA and AIA[2] to provide a function and systems oriented structure sensitive to the decisions required in early design (e.g., two stories vs. three stories, 40-ft vs. 30-ft spans, etc.).

The UNIFORMAT system provides various levels of estimating detail (Levels 1 through 8). Levels 6–8 are used for detailed estimating (quantity survey take-off) of completed design work. For budget development, the author recommends use of Level 3 to provide an adequate level of description. Appendix B provides the content of each UNIFORMAT code of accounts to Level 5.

Budget Estimate as a Scope Document

After calculation of space, configuration, and quantities and definition of quality, it is then the job of the cost engineer to apply the proper unit prices and add or delete items of work peculiar to the project.

In preparing a budget that represents a cost plan to be used to control costs during future design, it is important that several ground rules be established and followed.

The cost engineer plays a key role in defining project scope through the development of the budget estimate. The physical content of the estimate is an important scope document as well as explanations or drawings that may accompany it.

Scope is displayed in the estimate in the following fashion:

1. Through use of a standard code of accounts.

Use of a code of accounts accepted by the management of a particular organization can represent scope. Standard codes of accounts for estimates can indicate the content of work expected within each category if the published definition of the account is understood by all. For example:

- CSI accounts are understood in the construction industry to consist of a standard set of 16 specification divisions. The cost shown for each of the 16 divisions would influence the scope of work for that division.
- UNIFORMAT accounts are similarly understood to consist of 12 building systems with specified content.

2. Through documentation of criteria.

The criteria documented with each UNIFORMAT element must be used, along with the quantity and quality level, to establish the unit price for that parameter. For example:

- The ratio of open to closed space directly impacts the unit cost of partitioning, which is a function of the unit costs of fixed and moveable partitions.

3. Through inclusion.

The inclusion of systems and facility items in an estimate indicate scope. Conversely, the exclusion of systems and items in an estimate might indicate that they are not in the scope. For example:

- Not providing for a dual fuel system in 084, special mechanical systems, indicates that the system is not in the project.
- Omitting UNIFORMAT Account 07 from the estimate, or indicating it to be zero, would mean that the project had no conveying systems in its scope.
- Including an emergency generator in the budget estimate would indicate existence of an emergency power system in the project scope.

4. Through quantification.

The quantity shown for a particular system indicates the scope of the system. For example:

- The 64 FLT of stair construction on a 16-story building limits the number of stairwells to four, which in turn limits the floor size to the 50,000 square foot range.
- The KW size of the emergency generator would limit the amount of equipment connected to it.
- The square footage of the exterior wall would indicate the degree to which architectural expression is influenced through the ability to articulate the building's perimeter.

5. Through description.

The descriptions given each system or item in a budget estimate indicate the quality level represented by the unit prices used. This too, is scope because it defines range and limits the opportunity to change. For example:

- The material shown or generic system name indicates general item quality to be expected. For example, basing cost on precast concrete exterior walls would exclude less expensive stucco, would exclude more expensive stone or marble, but might include a comparable cost brick.
- Indicating copper piping would portray scope, whereas indicating just piping would not.
- Indicating a VAV air conditioning system, zoned every 3,000 square feet, would be far different in scope than just including an air conditioning system without further description.

6. Through pricing.

The budget level unit prices used to price each listed system and item also indicate scope within the price ranges known to be possible for each. For example:

- Exterior wall systems can cost between $12 and $50 per SF of wall. These prices represent a quality level from a simple, uninsulated metal curtain wall to fine stone or granite construction.
- Air conditioning can range from $1,200 to $7,500 per ton. The price used can infer the scope of the distribution system, type of plant equipment, the extent of humidity control, and the degree of temperature control.

DATA REFERENCES	
SYSTEMS COST DATA	
Means Assemblies Cost Data	- R. S. Means
Design Cost File	- Van Nostrand Reinhold
Dodge Systems Cost Manual	- McGraw Hill
Real Estate Valuation Guide	- Marshall & Swift
UNIT PRICE COST DATA	
Building Construction Cost Data	- R. S. Means
National Construction Estimator	- Craftsman
Berger Building Cost File	- Craftsman
Current Construction Costs	- Lee Saylor
Building Cost File	- Van Nostrand Reinhold
General Construction Estimating	- Richardson
OTHER REFERENCES	
The Building Estimators Reference Book	- Frank R. Walker

Figure 7.2 Cost data sources

Pricing Guidelines

Figure 7.2 provides a partial list of commonly used construction cost publications of value in the preparation of budget estimates for construction.

The best source, of course, is historical parameter data developed by the cost engineer. When pricing UNIFORMAT system quantities, follow these general rules:

1. Current commercial unit prices should be used.
2. Unit prices used should contain subcontractor markup only, and not contain general contractor's overhead or profit as that is shown separately in the UNIFORMAT system.
3. Unit prices should be adjusted to the current date of the estimate.

Using the Author's Database

Accompanying this text is the author's custom 2014 database. Readers can update this database and also add additional cost items in the database from their own experience or from some of the cost data sources listed in Figure 7.2.

Sample Estimate

The author prepared a sample estimate using his database for the Atlanta, Georgia, Office Building Headquarters project used as illustration in previous chapters. The database file provided (see Appendix A) is called data2014.xls. The sample BOE is based on only the following data:

- space program data—Chapter 3;
- configuration data—Chapter 4;
- computed quantities, systems 01–06—Chapter 5;
- plumbing model—Chapter 6;
- HVAC model—Chapter 6;
- electrical block load model—Chapter 6;
- computed quantities, systems 07–123—Chapter 6.

The sample estimate, provided at the end of this chapter, illustrates the qualitative descriptions for each system price selected by the author from the database. This represents a level of description of system scope.

Using the Cost Template

The sample estimate was prepared using the cost template file accompanying this book (see Appendix A). Its name is costtemplate.xls. This template is pre-formatted with the relevant UNIFORMAT section heads. Following the instruction on the template, one can copy lines from the database, paste them into the template, add the quantity, and then sort the template by ascending data to create their final estimate.

At any point one can add lines to the template, enter a new data line into the template from the database, and resort it to modify a previously prepared estimate.

Markups

Up to this point, we have discussed the development of direct construction costs without general contractor overhead and profit, contingency, escalation, or location factor adjustment. All costs in the database are priced at the subcontractor level.

Each of these cost elements must be added to the total direct subcontractor construction cost in the proper order of sequence. The cost template provides for the inclusion and computation of these cost elements in its last section.

Overhead and Profit

Overhead and profit must be added to the sum of direct construction cost and design contingency. This overhead and profit represents the general contractor's markup on the work of his subcontractors. Figure 7.3 provides guidance as to what markups will be charged based upon construction value and type of procurement.

A tight market exists when a lack of bidders is foreseen and competition is limited. Use the tight market column for design-build and turnkey projects. Competition does not really exist when it comes to construction cost.

Use the competitive market column if all project work is to be bid by the owner or construction manager. Use a percentage somewhere between the two columns if bidding is to take place among a select list of pre-qualified bidders. This is viewed as somewhat restricted competition.

Design Contingency

The method of budgeting discussed in this text performs quite well in summing the cost of those systems and criteria that are known or assumed. If one could ever be really dogmatic about cost control,

OVERHEAD & PROFIT (percent)		
WORK IN PLACE ECC VALUE (millions)	TIGHT MARKET	COMPETITIVE MARKET
<$1	17	13
$1.0–$2.5	15	11
$2.5–$5.0	14	10
$5.0–$10	13	9
$10–$20	12	8
$20–$50	11	7
>$50	10	6

Figure 7.3 Overhead and profit

then just those identical elements approved would be permitted in the design and everything else would be ruled out.

The big fear in this system of budget development from a developer, owner, or designer perspective is that something really necessary will be forgotten and left out of the budget. There are really only four solutions to correct such a discovery at a later date:

1. *Get more money.* Increase the budget. At least the budget document shows that the item or system needed was not included. An honest mistake was made.
2. *Leave it out.* Disapprove the discovered requirement. After all, it was not in the budget to begin with, and was therefore not approved and clearly not in the project scope.
3. *Make a trade-off.* Identify some other budget element or criterion and consciously change it to pay for the desired new requirement.
4. *Use budgeted contingency funds.* Add a reasonable design contingency. The design contingency would provide a source to fund things that might be forgotten. The amount of the design contingency should vary depending upon the size (value) of construction and knowledge of the developer, owner, designer, and the particular site.

Figure 7.4 provides a table of design contingency percentages to be added to the direct construction cost. This funding, even though its use is undesignated, will presumably represent increased construction cost and will disappear as soon a construction is awarded.

DESIGN CONTINGENCY (percent)		
WORK IN PLACE ECC VALUE (millions)	UNKNOWN SITE	KNOWN SITE
<$1	10	9
$1.0–$2.5	9	8
$2.5–$5.0	9	8
$5.0–$10	8	7
$10–$20	8	7
$20–$50	7	6
>$50	6	5

Figure 7.4 Design contingency

Do not confuse this with the construction contingency for change orders that is to be budgeted as discussed in the next chapter.

Escalation

If the job has been done right, the cost engineer has just predicted what the construction cost would be if the bids were taken tomorrow, because current unit prices were used.

Note that "current unit prices" reflect what would be bid today and, thus, contain enough cost in them to carry them to the midpoint of the future construction period. Prices based on historical bid data also contain midpoint cost.

Therefore, one needs only to escalate the budget between today's date and the proposed start of construction, which is one of the scheduling milestones explained earlier in this chapter.

Location Factor Adjustment

It takes some time to become experienced and familiar with a given market area. Normally, it is best for the cost engineer to base the budget on unit prices from the market area he is familiar with and then convert the final result to the market area where the project is to be located.

In the example used throughout this text, the author based pricing on the Washington, D.C. market area where he is located and with which he is familiar. The weighted average city cost index[3] for Washington, D.C. is 97.3. From the same source, the index for Atlanta, Georgia is 88.0. This means that approximately 10% must be deducted from the estimate based upon the author's Washington, D.C. prices to make it valid for Atlanta as computed below:

88.0 / 97.3 = 0.90.

This percentage is applied to overhead and profit as well as design contingency and escalation just calculated.

Check each overall parameter unit price to see if they meet your test of adequacy and if the overall cost per square foot is reasonable for the type and quality of space being budgeted. Do not accept the data just because it came from a computer. You control the input and can modify the output to satisfy your needs. This type of parameter estimating gives you another tool to use to check the construction portion of the budget you finally present.

For your information, the 2013 R.S. Means square foot cost for a mid range (five- to ten-) story building (without adjustment for location) as provided in three cost ranges is:

- low 1/4—$105.00;
- medium—$127.00;
- high 3/4—$172.00.

The following cost estimate using the system and database provided with this book comes out at $163.41 per square foot. Not too bad compared to the above data check.

CONSTRUCTION COST BUDGET

Project/Location: Atlanta, Georgia   Date:
Bldg./Const.Type: Office Building Headquarters   GSF: 274,549

SYSTEM	SUBSYSTEM	QTY	MEAS	UNIT COST	SYSTEM COST	$/GSF
01 Foundation						
Full system	Foundations—Mid rise 5–8 Story	51,500	FPA	36.83	1,896,951	6.91
02 Substructure						
Slab on grade	slab 5", LL 200 psf, storage areas	51,500	SFSA	5.38	276,864	1.01
Basement excavation	clay, 1 story deep	4,000	CY	11.39	45,570	0.17
Basement walls	CIP, 10 feet high, 10" thick	402	LF	128.61	51,703	0.19
Dampproofing	Hot asphalt, trowled on, primer plus 2 coat	4,824	SFSA	1.74	8,408	0.03
03 Superstructure						
Floor system	composite, LL 100 30x35 bay, 2–8 flrs	223,049	SSA	22.26	4,965,071	18.08
Stairs incl. handrails	CIP stairs, 44" wide	304	VFT	511.77	155,578	0.57
Roof construction	concrete, 30x35 bay	51,512	SSA	13.67	704,221	2.57
Canopy	concrete cantilevered	500	SSA	26.38	13,188	0.05
04 Exterior Closure						
Exterior wall	Precast, 4" U=0.10	62,230	SFSA	23.43	1,457,769	5.31
Exterior doors	Fire door, steel, single	4	EA	2970.35	11,881	0.04
Exterior doors	Entrance, aluminum, double	1	EA	4624.97	4,625	0.02
Exterior doors	Handicapped, alum-glass, automatic	1	EA	12785.56	12,786	0.05
Exterior doors	Overhead, steel rolling, manual, 12x12	2	EA	2392.98	4,786	0.02
Exterior doors	for motorized rolling door add	2	EA	925.96	1,852	0.01
Exterior windows	Fixed sash, aluminum, insl glass	26,670	SFSA	37.38	996,925	3.63
05 Roofing						
Full system	Membrane roof system	515	SQ	1233.92	635,468	2.31
Roof openings	Hatches, steel, factory primed	30	SFSA	96.47	2,894	0.01
06 Interior Construction						
Full system	Office space	203,850	NSF	29.87	6,089,509	22.18
07 Conveying System						
Passenger elevators	Hydraulic, 3500 lb. 4–7 floors	17	LO	26918.29	457,611	1.67
Freight elevators	Traction geared, 4000 lb 4–10 floors	5	LO	25663.10	128,316	0.47
081 Plumbing						
Full system	central core, 5–9 story	257	FU	4732.70	1,216,303	4.43
082 HVAC						
Heating	hot water, ducted AHU with HW coils	6,840	MBH	576.16	3,940,907	14.35
Ventilation	Ducted ventilation	9,000	NSF	1.76	15,876	0.06
Air Conditioning	VAV system, 1 zone/3500 sf	747	TON	5106.72	3,814,718	13.89
083 Fire Protection						
Full system	Halon	14,000	SFSA	22.21	310,905	1.13
Sprinklers	wet pipe system, ordinary hazard	274,549	SFSA	2.49	683,215	2.49
Fire Pumps	30 HP, 500 gpm	1	EA	18565.25	18,565	0.07
Standpipe System	wet risers, 4" Class I - 1 floor	4	STA	4305.00	17,220	0.06
Standpipe System	additional floors, 4"	16	STA	1255.19	20,083	0.07

Figure 7.5 Construction cost budget—sample project

084 Special Mechanical

Special system	Air curtain, loading dock, 8' x 5'	24	LF	444.22	10,661	0.04

09 Electrical

Full system	Mid rise electrical 5–9 story	4,500	AMP	1762.74	7,932,330	28.89
Special systems	Fire alarm system	274,549	GSF	1.14	314,221	1.14
Special systems	Security system, office space	274,549	GSF	0.19	51,890	0.19
Special systems	Telephone system, stub-up only	274,549	GSF	0.81	221,973	0.81
Emergency Power	Diesel generator, >400 kw	450	KW	352.23	158,505	0.58

11 Furnishings & Equipment

Equipment	Loading dock levellers	2	EA	19537.21	39,074	0.14
Equipment	Loading dock bumpers/boards	2	EA	7464.60	14,929	0.05
Directories	Boards, main building directory	1	EA	2833.57	2,834	0.01
Directories	Boards, floor directory	4	EA	1762.35	7,049	0.03
Window Treatment	Blinds, custom 1" slat	26,670	SFSA	9.27	247,271	0.90
Chutes	Mail chute, bronze/stainless	5	FLR	1347.18	6,736	0.02

12 Sitework

Site preparation	earth, flat—depth 1–3 ft	940,460	SFSA	0.6825	641,864	2.34
Site improvements	Flagpole, 20 ft. aluminum	2	EA	1111.152	2,222	0.01
Paving	Asphalt pavement for parking	1,230	CARS	1147.50	1,411,425	5.14
Paving	Plaza construction, granite pavers	2,000	SFSA	19.59	39,186	0.14
Site lighting	Parking lot lighting, 0.5 watts/sf	461,250	SFSA	0.86	397,136	1.45
Landscaping	Fine grade and seed	47,301	SY	2.32	109,762	0.40
Landscaping	Planting allowance	42,570	SFSA	0.92	39,335	0.14
Security Improvements	Fence, chain link, 10 ft	4,000	LF	17.99	71,946	0.26
Security Improvements	Security booth	1	EA	14977.47	14,977	0.05
Security Improvements	Gate—18 ft wide, sliding	1	EA	9447.32	9,447	0.03
Site Utilities	all utilities suburban site	940,460	SFSA	2.73	2,567,456	9.35
Subtotal Cost					42,271,997	**$ 153.97**
	Contractor OH & Profit	7.0	PCT		2,959,040	10.78
	Design Contingency	7.0	PCT		3,166,173	11.53
	Escalation	3.0	PCT		1,451,916	5.29
	Construction Index	90.0	PCT		(4,984,913)	-18.16
	Adjusted Total Construction Cost				44,864,213	**$ 163.41**

Legend

FPA	footprint area	SQ	100 square feet
SSA	supported surface area	GSF	gross square feet
NSF	net square feet	KW	1000 watts
VFT	vertical feet	LS	lump sum
FU	fixture units	STA	stations
LO	landing openings	EA	each
MBH	1000 Btuh	ACR	acres
TON	12,000 Btuh	PCT	percent

Figure 7.5 (continued)

The big advantage, however, is that you can see what is in the budget and what is not in the budget. The five pages of program detail as well as the description of the cost system items used in developing this budget now provide excellent tools for cost control of future design work.

Notes

1. GSA Handbook, *Capitalized Income Approach to Project Budgeting*, Washington, D.C.: U.S. General Services Administration.
2. Section B-5, Design & Construction Cost Management, *Architect's Handbook of Professional Practice*, vol. I., AIA, 1985, Washington, D.C.
3. R.S. Means, *Building Construction Cost Data*, 2013, Norwell, MA: Reed Construction LLC.

8   Soft Cost Determination

Introduction

Up to this point in the text the focus has been on the development of hard costs and the estimated cost of construction. This chapter deals with budgeting for all of the other costs shown in Figure 2.1, except for interest, which completes the total project cost. This begins with deciding on the project management plan to be used to accomplish the project.

Project Management Plan

The project management plan is a required element of project scope and is the second element of a design program (Figure 2.2). The plan is concerned with the management methods of design and construction to be used to meet the desired schedule. This plan should be developed before, or at a minimum concurrent with, budget development. The author recognizes that this is not normally done but suggests that this become the rule rather than the exception.

It is essential that the budget be based upon a project management plan intended to be followed to achieve the project. A change to the project management plan normally causes a change in the ultimate project cost. The budget amounts for design, construction, project administration, and escalation are all time related as is the expected income from a commercial development.

There are four issues involved in creating a project management plan; management organization, method of accomplishment, strategy, and schedule.

Management Organization

The plan should identify the developer or owner's organization to manage the design and construction of the project. This requires an assessment of technical capability, experience, and needs in reviewing and approving design and construction work.

If technical capability does not exist in-house, then it can be assumed that the developer will contract with consultants for the capability. Funds for contracting out these services need to be within the project budget. Funding for work to be performed by in-house staff might or might not be included in the project budget depending on the developer's policy. Key management tasks for which capability is needed include:

- Project management:

    ○ maintaining schedule;
    ○ overseeing all contract administration;
    ○ directing and redirecting;
    ○ approving progress payments.

- Design contract administration:

    ○ writing the A-E contract;
    ○ selecting the A-E;
    ○ negotiating the fee;
    ○ design supervision;
    ○ ensuring technical adequacy;
    ○ conducting value engineering study;
    ○ approving delivered work.

- Construction contract administration:

    ○ pre-qualifying construction firms;
    ○ writing the construction contract;
    ○ issuing construction contracts;
    ○ negotiating changes.

Methods of Accomplishment

There are many methods of accomplishing a project using one or more techniques such as design-build, turnkey, phased construction/fast tracking, construction management, or lump-sum bidding. Construction management can either be used itself as a method or be combined with any of the other methods.

Each of these methods has an influence on the quality, cost, and timeliness of providing the required facilities. To understand how,

DESIGN AND CONSTRUCTION METHODS	HIGH QUALITY	SHORT SCHEDULE	LOW FIRST COST	LOW LIFE CYCLE COST
• Conventional (Design then Build)	Best Chance	Worst Chance	Average Chance	Best Chance
• Fast Track (Overlapping D&C)	Average Chance	Average Chance	Best Chance	Average Chance
• Turnkey (Single D&C Contract)	Worst Chance	Best Chance	Worst Chance	Worst Chance

Figure 8.1 Methods of procurement

the methods of procurement (Figure 8.1) need categorizing and defining.

Conventional Method

The conventional method is a sequential process where design is completed first, then a single lump-sum competitively-bid contract is awarded to construct the facility.

- Quality. This method provides the best chance for highest quality. There is more time for design review and all design work is finished before construction starts.
- Cost. This method has highest risk for cost escalation.
- Schedule. Longest completion time.

Phased Method or Fast Tracking

The phased method begins as a sequential process where schematic design is completed first. Then the project is broken into several phases (bid packages) such as a package for foundations, shell, mechanical, electrical, interior finishes, and site work. Design is advanced and completed on the first package. The first package is placed under construction while design for the remaining packages continues.

- Quality. There is greater risk to quality because of less time for design review, resulting in greater chance for mistakes by starting construction before design is completed.

- Cost. Escalation cost is reduced; however, this is often offset by the need to hire a construction manager to coordinate site construction.
- Schedule. Time is reduced by as much as one-third of the time required for the conventional process.

Design-Build or Turnkey

This method places all the owner's contracting requirements at the beginning of the project. The owner contracts both for design and for construction at the same time. With single point responsibility for both services, project work is normally fast tracked.

- Quality. This method poses the greatest risk to the owner for poor quality.
- The owner has the greatest burden in contract preparation defining what is wanted and the greatest burden in reviewing what is delivered.
- Cost. Often, this is the cheapest initial cost solution because scope and quality are not adequately controlled.
- Schedule. This method should provide the fastest delivery.

Design Phases

The discussion above, regarding method of accomplishment, is often referred to as construction phases. This is not to be confused with the three normal design phases defined by the American Institute of Architects. The three design phases are normally referred to as the schematic, design development, and construction documents phases. All design work goes through the three design phases regardless of how construction is phased.

- Schematic phase.

Other organizations often refer to this first phase of design as the concept phase. Completed work at this stage should show intended massing, configuration, siting, elevations, single-line floor plans, and major building sections. Concept descriptions and block loadings for mechanical and electrical systems should be complete. Net and gross area computations should be complete. Models, renderings, and perspective drawings should be finished.

- Design development phase.

Other organizations often refer to this second phase of design as either the tentative stage of design or the 35 percent design stage. Design work from the schematic phase should be advanced. In addition, an outline specification should be complete. Structural, mechanical, and electrical design computations should be completed. If a project is to be divided into packages for phased construction or fast tracking, it is best that this not be done until after approval of the design development submission. Holding design work together at least until this phase is completed will help to reduce major errors, omissions, and causes for change orders.

- Construction documents phase.

Other organizations often refer to this third phase of design as the working drawing stage or 100 percent design stage. Completed plans and specifications for the entire project, or each bid package as appropriate, should be finished, ready for construction.

Recommended Strategy

The recommended method of procurement and project administration should be a function of the capability of the developer or owner's organization and their needs regarding the relative importance between quality, cost, and time.

If, however, a strategy cannot be elicited from the owner, the cost engineer must select a recommended procurement strategy based on the above. This strategy should be outlined in detail for all major steps in the process. Project phases, long-lead items and critical-path milestones should be identified. In addition, the cost engineer should include in the schedule the time it will take from the date the budget is submitted, to when it is funded, to set the date for design procurement.

Schedule

The second element required for definition of project scope is the establishment of a milestone schedule for the project, including budget approval, design, construction, equipment installation, testing, start-up, and full operation. Project phases, long-lead items, and critical-path milestones should be identified. The schedule should reflect the assumed specific method of acquisition as discussed above.

The milestone schedule will control project definition in terms of time. Cost will escalate over time in accordance with the labor and material rates in the geographical location. Impacts for delay caused by weather conditions on construction in the geographical location should be anticipated in the milestone schedule. One should use the milestone schedule to budget the cost of each element of the project in accordance with its planned commitment date.

Figure 8.2 provides typical completion timeframes for design work depending on its estimated construction value. The lower end of these time ranges should be used for projects that are simpler and less complex.

- Simpler projects generally meet the following criteria:

    ○   All one type of space is involved.
    ○   Floors are repetitive in nature.
    ○   The site is not congested.

- Complex projects generally meet the following criteria:

    ○   Multi-use space is involved.
    ○   The space is more technical in nature, i.e. hospitals, laboratories, etc.
    ○   The site is cramped.

DESIGN COMPLETION TIME		
ECC VALUE (millions)	SIMPLE PROJECT (months)	COMPLEX PROJECT (months)
<$1	4	6
$1–$5	6	8
$5–$25	8	10
$25–$50	12	12
Each additional $50	1	2

Figure 8.2 Design time

CONSTRUCTION COMPLETION TIME		
WORK IN PLACE ECC VALUE (millions)	SIMPLE PROJECT (months)	COMPLEX PROJECT (months)
< $1	5	7
$5	10	12
$10	12	18
$25	18	24
$40	21	26
$60	24	28
$100	30	36
$150	32	40
$400	50	60

Figure 8.3 Construction time

Figure 8.3 provides typical completion timeframes for construction work depending on its estimated construction value.

Soft Cost Worksheet

The author has provided a format with this text to assist in making soft cost computations (see Appendix A). The file name is softcost.xls. This file is actually a workbook containing a series of five worksheets appearing in the following order and titled:

- Soft Cost Worksheet;
- Payment Schedule Worksheet;
- Project Cash Flow;
- Final Project Budget;
- Rent Computation.

Illustrated next is Figure 8.4, the soft cost worksheet for the sample project. Again, one only needs to fill in the shaded cells on any of these worksheets to compute the budget data required.

The balance of this chapter will focus solely on the information necessary to complete the first four worksheets. The remaining worksheet, Rent Computation, will be discussed in the next chapter.

One begins the soft cost worksheet by filling in the estimated cost of construction as explained in the previous chapter. For the sample project, to start the soft cost worksheet, the author filled in the construction cost from the estimate created using the cost template as described in the previous chapter.

Soft Cost Worksheet

Project:	Office Building Headquarters	Date
Location:	Atlanta, Georgia	

Input Estimated Construction Cost (ECC) = $ 44,864,213

	Qty Meas.	Unit Cost	Subtotals	Totals
ECC—Additional Items				
Bond	0.5 % ECC		$ 224,321	
Insurance—Builder's Risk	1 LS	12,000.00	$ 12,000	
Demolition	1 LS	2,500.00	$ 2,500	
Contingency	5 % ECC		$ 2,243,211	
				$ 2,482,032
ESC—Estimated Site Cost				
Land	21 ACRE	55,000.00	$ 1,155,000	
Commissions	1 % LAND		$ 11,550	
Environmental Phase 1	1 STUDY	3,500.00	$ 3,500	
ALTA Survey	1 SURVEY	5,000.00	$ 5,000	
Soils Investigation	8 BORINGS	3,500.00	$ 28,000	
Title Work	1 LS	15,000.00	$ 15,000	
Appraisal	1 EACH	5,000.00	$ 5,000	
Real Estate Taxes	1 LS	50,000.00	$ 50,000	
Code/Zoning Analysis	1 EACH	5,000.00	$ 5,000	
				$ 1,278,050
EDC—Estimated Design Cost				
A-E Design	4.6 % ECC		$ 2,063,754	
A-E Reimbursables	5.0 % FEE		$ 103,188	
Observation Services	2 % ECC		$ 897,284	
As-Built Drawings	0.25 % ECC		$ 112,161	
Layout Drawings	1.0 % ECC		$ 448,642	
Special Studies	1 LS	20,000.00	$ 20,000	
Design Review	0.7 % ECC		$ 314,049	
VE Services	0.2 % ECC		$ 89,728	
Models	1 EA	5,000.00	$ 5,000	
Operating Manuals	0.25 % ECC		$ 112,161	
				$ 4,165,967
EMC—Estimated Management Cost				
Bid Proposal Expense	1 LS	25,000.00	$ 25,000	
CM Fee	3.5 % ECC		$ 1,570,247	
Local Fees	1 LS	15,000.00	$ 15,000	
Permits	1 LS	75,000.00	$ 75,000	
Developer Overhead	36 MONTHS	4,500.00	$ 162,000	
Reimbursables	5.0 % DEV OH		$ 8,100	
Testing	1 LS	8,000.00	$ 8,000	
Inspecting Architect	24 MONTHS	3,000.00	$ 72,000	
Legal	120 HOURS	250.00	$ 30,000	
Insurance	1 LS	10,000.00	$ 10,000	
Consultant	1 EA	15,000.00	$ 15,000	
Facility Management Prep.	2 MOS	4,000.00	$ 8,000	
				$ 1,998,347
EFC—Partial Financing Cost				
Financing Fee	0.5 % LOAN	$ 60,500,000	$ 302,500	
			Total	$ 10,226,896

Figure 8.4 Soft cost worksheet

ECC Additional Items

Four items that are not included in the construction estimate, which are normally associated with construction costs, need to be separately identified for lender financing. These are: bond, builder's risk insurance, demolition, and change order contingency.

Bond

Performance and payment bonding is normally required for lender financing. Some developers and owners do not include bond because they count on the size, net worth, and reputation of a large contractor. Often, when this occurs, the contracts are not competitively bid. Bond cost is normally budgeted as a percentage of construction cost as shown in Figure 8.5 or one can call a bonding company and get an exact quote.

Builder's Risk Insurance

Insurance during construction should be quoted by an insurance agent based on the class of construction, its estimated cost, and its duration. Some points to consider:

- Be prepared to specify if you want a deductible or not. For a $25,000 deductible, rates can be reduced 13–34 percent.
- "All risk" insurance excludes floods, earthquakes, and other perils. Consider your geographical location and risk. Some lenders are not risk tolerant.

PERFORMANCE BONDS	
ECC AMOUNT (millions)	PERCENTAGE
<$1	2.5
$1–$2	1.7
$2–$4	1.1
$4–$8	0.9
$8–$10	0.8
$10–$20	0.7
$20–$50	0.6
>$50	0.5

Figure 8.5 Bond percentages

Demolition

Some place this in construction cost. Others place this in site cost. So it is not forgotten, the author calls it out as an additional item. When calculating demolition some points to consider are:

- closure costs if underground tanks are involved;
- containment costs if asbestos materials are involved;
- distance of dumping locations.

Change Order Contingency

Change order contingency is that amount budgeted to pay for design errors and omissions (contractor claims) during construction. Normally, the amount of change order contingency acceptable to a lender is in the range of 3.5 percent to 5 percent of estimated construction cost. The lower percentage is often used for commercial tenants, the higher percentage for government tenants, or unique buildings.

Estimated Site Cost (ESC)

Site cost consists of the cost of the land as well as cost for all related due diligence activities a developer would normally perform for a lender prior to closing. As shown in Figure 8.4 these include: commissions, environmental report, ALTA survey, preliminary soils investigation, title work, appraisal, real estate taxes during the construction period, and code and zoning analysis.

Estimated Design Cost (EDC)

The budget for EDC includes all design consulting service costs for the project. Figure 8.6 provides recommended percentage amounts to budget for design work, observation (inspection) service, and design review (often by a construction manager or independent third party). These percentages are multiplied against the ECC amount. In addition to the above percentages, one will want to consider budgeting for the following items, which vary according to the specific project.

A-E Reimbursables

Reimbursable expenses will normally range from 5–10 percent of the A-E design fee. The major cost items included in reimbursable expenses are:

DESIGN SERVICES			
WORK IN PLACE ECC VALUE (millions)	DESIGN WORK	OBSERVATION SERVICE	DESIGN REVIEW
<$1	7.7	7	5.1
$1.0–2.5	7	5.2	3.6
$2.5–5.0	6.4	4.3	2.5
$5.0–10.0	5.9	3.9	1.5
$10.0–20.0	5.4	3.5	1
$20.0–40.0	4.9	2.8	0.7
$40.0–60.0	4.6	2	0.6
$60.0–90.0	4.4	1.6	0.5
>$90.0	4.1	1.4	0.4

Figure 8.6 Design percentages

- travel cost if designer, client, and project are not located in same city;
- reproduction of plans and specifications for submittals and bidding;
- shipping expenses for documents.

Other Design Phase Cost

The following additional cost items shown on Figure 8.4 are included in EDC:

- As-built drawings. Normal cost range—0.25 to 0.75 percent.
- Layout drawing service. Normal cost range—1.0 to 1.5 percent.
- Special studies—depends on need for special consultants to deal with special program items like acoustics, cafeteria, elevator, security, etc. Normal cost range is $10,000 to $25,000 per issue.
- Value engineering services. Normal cost range—0.1 to 0.5 percent.
- Operating manuals. Normal cost range—0.25–0.75 percent.

Estimated Management Cost (EMC)

Management costs are the remaining potpourri of expenses incurred by a developer or owner during the life of a project. The list shown in Figure 8.4 includes the following.

Bid proposal expense

If desired, this is the place to budget for the recovery of any out-of-pocket direct cost incurred in winning the procurement.

CM fee

It is normally not economical to use the CM method of procurement on projects less than $10 million in value. If the CM is employed early to provide guidance during design (which is highly recommended) approximately one-third of the CM fee can be assigned to the EDC cost account under the design review item. The total CM fee should not exceed 3.0 to 3.5 percent of ECC.

Developer overhead

Developer overhead is normally budgeted on a monthly basis for the duration of the project from the first day of financing through close-out. Developer overhead includes, at a minimum, salaries for project staffing, home office and field expense, and travel. Total overhead should not exceed 2–3 percent of ECC.

Inspecting architect

Most lenders will want the developer to pay the cost of their inspecting architect. This cost will be incurred monthly, one month prior to start of construction and through the entire construction period. Normal budget range for this item is $2,500–$3,500 per month.

Other expenses

Other expenses shown in Figure 8.4 include:

- Local fees. Check the jurisdiction.
- Permits. Check the jurisdiction.
- Reimbursables. Like the A-E, developer reimbursables should be 5–10 percent of developer overhead.
- Testing. The major testing costs will be for soil compaction, concrete strength, and bolting torsion for steel. A concrete structure will cost more to test than will a steel frame structure.

- ◦ For concrete, normal budget range is $6,000 to $7,500 per story.
- ◦ For steel, normal budget range is $5,000 to $6,000 per story.
- Legal. Get a quote from your attorney for loan document review and closing cost.
- Insurance. This is for liability insurance for the team during the design and construction period. Get a quote from your agent.

Facility Management Preparation

For a major new building one should staff the building manager on site at least two months prior to completion of construction and budget for salary and expenses.

This individual should become familiar with the building design, interface with subcontractors to administer future warranty work, observe manufacturer start-up of each piece of new equipment, receive testing reports, receive manufacturer training in operating all equipment, and receive all operating manuals. In addition, this individual should prepare all vendor service contracts for janitorial, landscaping, elevator maintenance, trash removal, etc.

Other

Other budget amounts may need to be held in reserve by the owner for a wide variety of reasons unique to the project. Following is a checklist of these expenses the author has incurred on previous projects for budget consideration:

- building artwork—federal government budgets 0.5 percent;
- moving expenses of tenants;
- temporary rental of space for tenants;
- partial or early occupancy expense;
- telephone instruments and their installation;
- owner furnished equipment to be contractor installed;
- occupant loose furniture and furnishings;
- environmental costs, i.e. archeological exploration, relocation of graves, protection of trees;
- historic preservation;
- relocation, i.e. cost to move and resettle existing residents and businesses located on a site: buying them out or paying for their rent elsewhere.

Estimated Financing Cost (EFC)

Financing cost includes three things; the financing fee to secure the loan, interest on the loan during development, and the interest reserve desired by the lender to hedge a change in rate or a change in the anticipated draw schedule.

Financing fee

Financing fee is the up-front points your lender will charge to make the loan. The normal range is 0.5–2.0 percent. The important point to remember is that this percentage is on the full value of the loan, which includes all construction costs and all soft costs (including the financing fee and interest on the loan).

   This is a circular type of computation because you must know the loan amount (the answer) to compute the fee. However, with the author's format worksheets it is easy. To estimate financing fee:

1. Guess at the total loan amount for the soft cost worksheet (Figure 8.4).
2. Enter the financing fee percentage.
3. Finish your computations (payment schedule and cash flow schedule).
4. Obtain the resulting loan amount from the final budget worksheet.
5. Go back and revise the loan amount on the soft cost worksheet and recalculate the budget cash flow schedule.

Interest

Accurate budgeting for construction interest is essential since it is the second biggest cost item after construction cost and requires management prowess to control. Interest cost can make or break a deal.

   Control of interest cost requires a well documented basis for computing it that can be understood, monitored, and redirected before it overruns. The author's format tool, softcost.xls makes this much easier. The second worksheet of this file is the Payment Schedule Worksheet (Figure 8.7).

Payment Schedule Worksheet

Project: Office Building Headquarters   **Date** _____

Location: Atlanta, Georgia

Project Start Month	0
Construction Period	24
Month Construction Start	12
Project End Month	36

(Must be 36 or less)

ECC—Additional Items			Month to be Paid
Bond	$	224,321	0
Insurance—Builder's Risk	$	12,000	0
Demolition	$	2,500	1
Contingency	$	2,243,211	36
ESC—Estimated Site Cost			
Land	$	1,155,000	0
Commissions	$	11,550	0
Environmental Phase 1	$	3,500	0
ALTA Survey	$	5,000	0
Soils Investigation	$	28,000	0
Title Work	$	15,000	0
Appraisal	$	5,000	0
Real Estate Taxes	$	50,000	12
Code/Zoning Analysis	$	5,000	0
EDC—Estimated Design Cost			
A-E Design	$	2,063,754	1–12
A-E Reimbursables	$	103,188	1–12
Observation Services	$	897,284	13–36
As-Built Drawings	$	112,161	36
Layout Drawings	$	448,642	9–12
Special Studies	$	20,000	4
Design Review	$	314,049	2–8
VE Services	$	89,728	3,6
Models	$	5,000	0
Operating Manuals	$	112,161	36
EMC—Estimated Management Cost			
Bid Proposal Expense	$	25,000	0
CM Fee	$	1,570,247	9–36
Local Fees	$	15,000	12
Permits	$	75,000	12
Developer Overhead	$	162,000	1–36
Reimbursables	$	8,100	1–36
Testing	$	8,000	13,14,15
Inspecting Architect	$	72,000	13–36
Legal	$	30,000	0
Insurance	$	10,000	0
Consultant	$	15,000	3
Facility Management Prep.	$	8,000	35,36
EFC—Estimated Financing Cost			
Financing Fee	$	302,500	0
	$	**10,226,896**	

Figure 8.7 Payment schedule

This worksheet provides for the integration of project schedule with the expected payment of cost.

- The worksheet is structured to provide for a maximum 36 month schedule for both design and construction.
- For projects longer than 36 months, one would have to improvise by using two workbooks and splitting the cost between them.

Payment Schedule Worksheet Instructions

Cost items from the Soft Cost Worksheet are automatically brought forward to the Payment Schedule Worksheet.

1. At the top of the worksheet enter the total number of months necessary to construct the project. Clicking on the space causes a drop-down box to appear. The choices for construction duration that are preprogrammed are 6, 9, 12, 15, 18, 21, and 24 months.
2. Next, enter the month that construction is expected to start.
3. Review the Total Project Months to ensure that it does not exceed 36. If so, adjust either the construction duration or construction start month.
4. Next, enter the month of payment for each cost item in the payment column.

    - There are three forms of entry:
        - single number, i.e. 2—for full payment in a single month #2;
        - multiple payment months, i.e. 2,4,7,11—for full payment divided among those four months;
        - monthly payments, i.e. 7–24—for full payment divided among those months.

    - No month number can exceed the total number of months used.
    - Month number zero (0) can be used. It means the initial cash payment made at the time of financial closing. This is when interest starts.

Note the entry format used (numbers, commas, and dashes with no spaces) in the sample project payment schedule worksheet, Figure 8.7.

Project Cash Flow

Project cash flow (sometimes referred to as the draw schedule) is the third worksheet in the softcost.xls series. See Figure 8.8. The computer performs most of the work for this worksheet. All you have to do is enter two pieces of data:

1. Enter the annual interest rate as a percent, as a whole number.
2. Enter the construction payment retention desired as a percent, as a whole number.
3. Press the "calculate" button.

Project Cash Flow

Project: Office Building Headquarters   Date: _____
Location: Atlanta, Georgia

Annual Interest   6.00 %   Interest Rate = 0.005000 %
Const. Retention   10.00 %

Calculate costs   Review Weights

Month	Construction Cost	Design Fees	Other Soft Costs	Cumulative Payments	Monthly Interest	Monthly Cash Flow
0	0	0	1,836,871	1,836,871	9,184	1,846,055
1	0	180,578	7,225	2,024,675	10,123	197,927
2	0	180,578	49,589	2,254,842	11,274	241,442
3	0	180,578	109,453	2,544,874	12,724	302,756
4	0	180,578	69,589	2,795,042	13,975	264,143
5	0	180,578	49,589	3,025,209	15,126	245,294
6	0	180,578	94,453	3,300,241	16,501	291,533
7	0	180,578	49,589	3,530,409	17,652	247,820
8	0	180,578	49,589	3,760,577	18,803	248,971
9	0	180,578	172,966	4,114,121	20,571	374,115
10	0	180,578	172,966	4,467,665	22,338	375,883
11	0	180,578	172,966	4,821,209	24,106	377,650
12	403,778	180,578	312,966	5,718,532	28,593	925,915
13	605,667	37,387	66,472	6,428,057	32,140	741,666
14	807,556	37,387	66,472	7,339,472	36,697	948,112
15	1,009,445	37,387	66,472	8,452,775	42,264	1,155,567
16	1,211,334	37,387	63,805	9,765,301	48,827	1,361,352
17	1,413,223	37,387	63,805	11,279,716	56,399	1,570,813
18	1,615,112	37,387	63,805	12,996,020	64,980	1,781,284
19	2,018,890	37,387	63,805	15,116,102	75,581	2,195,662
20	2,422,668	37,387	63,805	17,639,961	88,200	2,612,059
21	2,826,445	37,387	63,805	20,567,599	102,838	3,030,476
22	3,028,334	37,387	63,805	23,697,125	118,486	3,248,012
23	3,028,334	37,387	63,805	26,826,652	134,133	3,263,660
24	2,826,445	37,387	63,805	29,754,289	148,771	3,076,409
25	2,624,556	37,387	63,805	32,480,038	162,400	2,888,149
26	2,422,668	37,387	63,805	35,003,897	175,019	2,698,879
27	2,220,779	37,387	63,805	37,325,868	186,629	2,508,600
28	2,018,890	37,387	63,805	39,445,950	197,230	2,317,311
29	1,817,001	37,387	63,805	41,364,143	206,821	2,125,013
30	1,615,112	37,387	63,805	43,080,446	215,402	1,931,706
31	1,413,223	37,387	63,805	44,594,861	222,974	1,737,389
32	1,211,334	37,387	63,805	45,907,387	229,537	1,542,063
33	807,556	37,387	63,805	46,816,135	234,081	1,142,829
34	605,667	37,387	63,805	47,522,994	237,615	944,474
35	403,778	37,387	67,805	48,031,964	240,160	749,130
36	4,486,421	37,387	2,535,337	55,091,109	275,456	7,334,601
	$ 44,864,213	$ 3,064,226	$ 7,162,670		$ 3,753,611	$ 58,844,720

Figure 8.8 Cash flow

Final Project Budget   Date _____

Project: Office Building Headquarters
Location: Atlanta, Georgia

		Subtotals		Totals
ECC—Estimated Cost of Construction				
Construction Cost	$	44,864,213		
Bond	$	224,321		
Insurance—Builder's Risk	$	12,000		
Demolition	$	2,500		
Contingency	$	2,243,211		
			$	47,346,245
ESC—Estimated Site Cost				
Land	$	1,155,000		
Commissions	$	11,550		
Environmental Phase 1	$	3,500		
ALTA Survey	$	5,000		
Soils Investigation	$	28,000		
Title Work	$	15,000		
Appraisal	$	5,000		
Real Estate Taxes	$	50,000		
Code/Zoning Analysis	$	5,000		
			$	1,278,050
EDC—Estimated Design Cost				
A-E Design	$	2,063,754		
A-E Reimbursables	$	103,188		
Observation Services	$	897,284		
As-Built Drawings	$	112,161		
Layout Drawings	$	448,642		
Special Studies	$	20,000		
Design Review	$	314,049		
VE Services	$	89,728		
Models	$	5,000		
Operating Manuals	$	112,161		
			$	4,165,967
EMC—Estimated Management Cost				
Bid Proposal Expense	$	25,000		
CM Fee	$	1,570,247		
Local Fees	$	15,000		
Permits	$	75,000		
Developer Overhead	$	162,000		
Reimbursables	$	8,100		
Testing	$	8,000		
Inspecting Architect	$	72,000		
Legal	$	30,000		
Insurance	$	10,000		
Consultant	$	15,000		
Facility Management Prep.	$	8,000		
			$	1,998,347
EFC—Partial Financing Cost				
Financing Fee	$	302,500		
Interest	$	3,753,611		
Interest Reserve	$	187,681		
			$	4,243,791
			$	59,032,400
Developer Fee/Profit	2.5 %		$	1,475,810
		Total Loan Amount	$	60,508,210

Figure 8.9 Final project budget

The computer performs the following tasks:

- distributes the ECC to the construction cost column in accordance with a preset bell distribution curve based on author experience;
- distributes the first four EDC costs to the design fees column based upon the months input on the Payment Schedule Worksheet;
- distributes all other soft costs to the soft cost column based upon the months input on the Payment Schedule Worksheet.

The computer calculates the monthly interest, the monthly cash flow, and the cumulative payments. The monthly interest payment is added to the cash flow and cumulative payment because it too must be borrowed in the loan.

Final Project Budget

The fourth worksheet of softcost.xls is the final project budget as illustrated by Figure 8.9.

The computer carries forward all previous data and computations including the project interest cost just computed. Now there are only two things left to do to complete the budget:

1. Enter the interest reserve amount.

   - Normally lenders will accept an interest reserve that is 10 percent of the calculated interest based upon the cash flow schedule.
   - An alternate approach would be to go back to the Cash Flow Worksheet and increase the interest rate by 1 percent and then enter the computed interest difference between the two rates as the interest reserve amount.

2. Enter the developer fee/profit percentage

Financing fee check

Now that you know the loan amount, go back to the first worksheet, Soft Cost Worksheet in softcost.xls, and at the bottom enter the loan amount to recompute the financing fee. Hit "calculate" again on the Cash Flow Worksheet and you now have the final loan amount.

9   Operating Cost and Rent

Introduction

Developers competing to provide space in the market place need to know how much rent they will have to charge before they get too far into the project. The budgeting effort explained to this point is a minimal expense compared to engaging a full design team, and it is more accurate than simply guessing at cost per square foot.

If the budgeted project is not competitive, the developer has full documentation containing a macro level of detail regarding project scope, criteria, and content to review and manipulate until an acceptable result is achieved.

This chapter provides and discusses the final tool provided by the author for the conversion of expected total project cost into full service rent. If one can compute full service rent, then triple net rent (or some other variation) is easily determined by leaving out some of the costs.

The Annual Cost Format

Full service rent includes not only return of capital (the project budget cost) but also the cost of the annual cost of operation and maintenance (O&M) during the rental period.

The author has provided a set of three computer worksheets to assist in the computation of annual cost (see Appendix A). The file name for these worksheets is annualcost.xls. The three worksheets in this file are called:

- CAP-X—capital expenses;
- O&M—unit cost expenses;
- 1217—Lessor's Annual Cost statement (based on GSA Form 1217).

One only needs to fill in the highlighted data on each worksheet as explained below for each worksheet.

Capital Expense Computation

Figure 9.1 below illustrates the capital expense computation (CAP-X) worksheet for the sample project.

Capital expenses are commonly estimated on a cost per square foot basis and are commonly referred to as reserves for replacement. The author has seen this cost estimated by best guess from between $0.15–$0.35 per square foot. Developers who plan to sell the property will shave this figure on the low side to win the bid. Owners who want to securitize their loan want a rental stream of income that will provide ample reserves to meet all lease requirements.

The best way to determine how much to include in the rental computation is to estimate replacement needs on a methodical basis. Figure 9.1 is organized to follow the UNIFORMAT code of accounts. It displays under each account those system components most likely to require replacement during the life of the facility.

Capital Expense Worksheet
Building Maintenance and Reserves for Replacement (less than 20 years)

Project: Office Building Headquarters
Location: Atlanta, Georgia

Category Item	Quantity	Meas.	Unit Cost	Replacement Cost	Life (yrs)	Annualized Amount
04 Exterior Closure						
Cleaning	62,230	SFSA	0.20	12,446	15	$ 830
Joint Repair	10,370	LF	1.25	12,963	15	$ 864
05 Roof						
Reroofing	515	SQ	950.00	489,250	12	$ 40,771
06 Interior Construction						
Carpet replacement	21,000	SY	27.00	567,000	10	$ 56,700
Painting	108,000	SFSA	0.80	86,400	3	$ 28,800
Wall covering replacement	185,000	SFSA	2.25	416,250	15	$ 27,750
Tile repair	3,000	SF	3.50	10,500	10	$ 1,050
081 Plumbing						
HW heater replacement	5	EA	1,200.00	6,000	12	$ 500
082 HVAC						
AHU fans	10	EA	850.00	8,500	15	$ 567
Package unit compressors	15	TON	950.00	14,250	8	$ 1,781
092 Lighting & Power						
Ballast replacement	2,030	EA	35.00	71,050	6	$ 11,842
12 Sitework						
Pavement restriping	1,230	CARS	9.00	11,070	3	$ 3,690
Pavement seal coating	399,750	SF	0.15	59,963	9	$ 6,663
Pavement overlay	399,750	SF	1.75	699,563	12	$ 58,297
				Annual Reserve Amount	$	240,104

Figure 9.1 Capital expenses for sample project

For budgeting simplicity that is accurate for its need, these costs are calculated as follows:

- The capital expense computation shown uses the constant dollar approach to computing lifecycle cost. Following this approach one estimates all costs based on unit prices for the present year.
- Inflation of cost for future years is ignored as is the interest to be accumulated when funds are placed in the maintenance reserve account.
- The life span (in years) indicated for each system component is the expected replacement year.
- The annualized amount is the computed replacement cost divided by the life.

The procedure to follow to estimate CAP-X expenses is:

1. Identify the significant replacement components based on the program requirements just budgeted, the design life of system components, and/or the requirements of GSA's SFO lease (if any). For example:

   - Most GSA leases require carpet replacement every ten years whether it is worn out or not.
   - Most GSA leases require repainting of public space every three years and repainting of the rest of the building every five years.

2. Unprotect the CAP-X format (Figure 9.1) to add additional estimating lines to include special components as necessary. Do not forget to copy down the annualized amount formula to these new lines so that they are included in the total annual reserve amount.
3. Determine the quantity of each component based on quantities computed in accordance with Chapters 5 and 6.
4. Determine unit prices based upon one of the references in Figure 7.2.

Unit Cost Expenses

The second worksheet under the annualcost.xls series is the Operations and Maintenance Cost (O&M) worksheet. This worksheet is illustrated as Figure 9.2. It portrays the cost per net square foot to operate and maintain the facility and it too is based on current cost.

Operations & Maintenance Cost Worksheet

Project: **Office Building Headquarters**   NSF = 203,850
Location: **Atlanta, Georgia**

		$/NSF
Custodial operations		
	Janitorial	$ 0.79
	Janitorial supplies	$ 0.11
	Toilet supplies (Soap, towels, tissues)	$ –
	Window washing	$ 0.03
	Trash removal	$ 0.06
	Snow removal	$ –
Maintenance and repair		
	Heating system	$ 0.08
	Air conditioning system	$ 0.17
	Electrical system	$ 0.06
	Electrical ballast, tube replacement	$ 0.03
	Plumbing system	$ 0.03
	Elevators	$ 0.10
	Lawn & landscape	$ 0.22
Other	Emergency generator	$ 0.03

		$/NSF		Consumption	Measure	$ rate
Utilities						
	Heating—oil	$ –			gallons	
	Heating—gas	$ –			ccf	
	Heating—electric	$ 0.93		3,464,992	kwh	$ 0.055
	Air conditioning—electric	$ 1.35		5,010,481	kwh	$ 0.055
	Electric—light and power	$ 0.63		2,351,750	kwh	$ 0.055
	Electric—special equipment	$ –			kwh	
	Water and sewer	$ 0.17				
Management						
	Building engineer and/or manager	$ 0.37				
	Security (Watchman, guards, not janitors)	$ 0.40				
	Social Security Tax / Workmans Comp.	$ 0.06				
Fixed Expenses				**Totals**		
	Real estate tax	$ 1.53		Variable	$	5.63
	Insurance	$ 0.14		Fixed	$	3.85
	Reserves for replacement	$ 1.18			$	9.48
	Property management	$ 1.00				

Figure 9.2 O&M costs for sample project

Operating Expense

Operating expenses are those expenses caused by use of a facility. The first four categories shown in Figure 9.2 are commonly referred to as operating expenses.

These are expenses for custodial, maintenance and repair, utilities, and management. Begin the O&M worksheet by entering the net square feet (NSF) at the top of the page.

Fixed Expenses

Fixed expenses are those expenses to own and manage a property even if it is not used or occupied. Such expenses include taxes, insurance, reserves for replacement, and property management as shown by the last category in Figure 9.2.

A good source of data for expected income and expense for office type space is published by the Building Owners and Managers Association (BOMA). Figures 9.3a through 9.3e provide a sample of the type of BOMA[1] data available.

1. From the BOMA data the author selected the city—Atlanta— the location—suburban—and the building size range— 100,000–299,999 square feet. This produced the following charts:

Income and Expense Overview–2011									
	Total Building Rentable Area				Total Office Rentable Area				
39 Bids	6,053,123 Sq. Ft.				6,016,012 Sq. Ft.				
	Dollars/S.F.		Mid Range		Dollars/S.F.		Mid Range		
# Bids	Avg	Mdn	Low	High	Avg	Mdn	Low	High	
Income									
Total Rental Income	37	19.44	18.13	15.52	23.92	19.57	18.41	15.52	23.92
Total Income	37	20.00	18.15	15.90	23.92	20.13	18.43	15.90	23.92
Expense									
Total Oper Exp	39	5.72	5.62	4.98	6.28	5.75	5.62	5.03	6.32
Total Oper + Fixed Exp	38	7.59	7.74	6.89	8.63	7.63	7.75	6.89	8.77
Income and Expense Summary–2011									
Income									
Office Rent	37					18.79	17.88	15.52	22.48
Retail Rent									
Other Rent									
Telecom Income									
Miscellaneous Income	21	1.08	0.37	0.04	0.83				
Expense									
Cleaning	38	0.87	0.88	0.75	1.01	0.87	0.89	0.75	1.01
Repair / Maintenance	38	1.06	0.86	0.54	1.37	1.07	0.86	0.55	1.37
Utility	37	1.98	1.96	1.48	2.26	2.00	1.96	1.48	2.26
Roads / Grounds	38	0.26	0.26	0.22	0.32	0.26	0.26	0.22	0.32
Security	35	0.40	0.19	0.10	0.45	0.40	0.19	0.10	0.46
Administrative	38	1.36	1.28	1.00	1.87	1.37	1.28	1.00	1.90
Fixed	36	1.93	1.84	1.53	2.68	1.94	1.84	1.53	2.68
Directly Expensed Leasing	18	3.71	1.89	0.37	5.24	3.72	1.89	0.37	5.34
Amortized Leasing	10	1.15	0.91	0.27	1.89	1.16	0.91	0.27	1.89

Figure 9.3a Atlanta occupancy summary

Income and Expense Detail–2011									
		Total Building Rentable Area				Total Office Rentable Area			
		Dollars/S.F.		Mid Range		Dollars/S.F.		Mid Range	
	# Bids	Avg	Mdn	Low	High	Avg	Mdn	Low	High
Income									
Office Rent									
Base Rent	37					17.24	17.50	14.85	20.55
Pass Throughs	25					2.59	0.81	0.37	3.72
Escalations	13					1.08	0.88	0.42	1.34
Lease Cancellations									
Rent Abatements (−)	19					1.29	0.98	0.28	2.34
Telecom Income									
Rooftop Income									
Wire/Riser Access Income									
Miscellaneous Income									
Gross Parking Income	7	2.87	3.66						
Tenant Service Income	11	0.29	0.26	0.07	0.47				
Other Misc. Income	11	0.16	0.03	0.01	0.13				
Expense									
Cleaning									
Payroll, Taxes, Fringes									
Routine Contracts	30	0.57	0.59	0.50	0.69	0.57	0.59	0.50	0.69
Window Washing	28	0.03	0.02	0.01	0.03	0.03	0.02	0.01	0.03
Other Specialized Contracts	13	0.08	0.07	0.01	0.13	0.08	0.07	0.01	0.13

Figure 9.3b Atlanta income and expense detail

2. The next step is to obtain the following three income and expense detail sheets for the subject property location.

Utilities

As shown in Figure 9.2, the author estimates utilities based on consumption and rate because this best ensures representing the uniqueness of the building being planned with respect to its program and technical criteria, and anticipated operating profile. One can always check the computed result against BOMA data to see if the computed cost is in line with the historical cost.

Supplies / Materials	28	0.11	0.10	0.07	0.14	0.11	0.10	0.07	0.14
Trash Removal / Recycling	30	0.06	0.05	0.03	0.08	0.06	0.05	0.03	0.08
Miscellaneous / Other	10	0.03	0.02	0.01	0.02	0.03	0.02	0.01	0.02
Repair / Maintenance									
Payroll, Taxes, Fringes	28	0.42	0.44	0.32	0.50	0.43	0.44	0.32	0.51
Elevator	25	0.10	0.08	0.06	0.14	0.10	0.08	0.06	0.14
HVAC	30	0.25	0.19	0.09	0.33	0.25	0.19	0.09	0.33
Electrical	29	0.06	0.05	0.02	0.08	0.06	0.05	0.02	0.08
Structural / Roofing	20	0.04	0.01	0.00	0.03	0.04	0.01	0.00	0.03
Plumbing	29	0.03	0.02	0.01	0.04	0.03	0.02	0.01	0.04
Fire / Life Safety	27	0.07	0.06	0.04	0.09	0.07	0.06	0.04	0.09
General Building Interior	26	0.20	0.13	0.07	0.29	0.20	0.13	0.07	0.29
General Building Exterior	12	0.02	0.01	0.01	0.02	0.02	0.01	0.01	0.02
Parking Lot	22	0.15	0.03	0.02	0.15	0.15	0.03	0.02	0.15
Miscellaneous / Other	18	0.09	0.05	0.02	0.10	0.09	0.05	0.02	0.10
Utility									
Total Electricity	29	1.89	1.81	1.45	2.15	1.90	1.81	1.45	2.15
Gas									
Fuel Oil									
Steam									
Chilled Water									
Water / Sewer	29	0.17	0.16	0.11	0.21	0.17	0.16	0.11	0.21
Roads / Grounds									
Landscaping	28	0.21	0.22	0.17	0.26	0.22	0.22	0.17	0.26
Snow Removal	19	0.03	0.02	0.01	0.03	0.03	0.02	0.01	0.03
Miscellaneous / Other	13	0.07	0.03	0.02	0.10	0.07	0.03	0.02	0.10

Figure 9.3c Atlanta income and expense detail (*continued*)

Consumption for air conditioning and heating load is taken from the HVAC model (Figure 6.5).

Consumption for lighting and power is taken from the Electrical Block Load computation (Figure 6.7). Use the demand load KW (less the HVAC demand load) multiplied by the normal operating hours for the building. For the sample project this is:

2,556.1 kw – (1,494 + 121.4) × 250 days/yr × 10 hrs/day = 2,351,750 kwh

Security									
Payroll, Taxes, Fringes									
Contracts	23	0.52	0.34	0.20	0.82	0.52	0.34	0.20	0.82
Equipment	10	0.02	0.01	0.01	0.03	0.02	0.01	0.01	0.03
Miscellaneous / Other	13	0.02	0.02	0.00	0.04	0.02	0.02	0.00	0.04
Administrative									
Payroll, Taxes, Fringes	25	0.48	0.39	0.34	0.49	0.48	0.39	0.34	0.49
Management Fees	27	0.53	0.54	0.45	0.65	0.53	0.55	0.45	0.65
Professional Fees	11	0.06	0.05	0.01	0.08	0.06	0.05	0.01	0.08
General Office Expenses	26	0.17	0.14	0.08	0.23	0.17	0.14	0.08	0.23
Employee Expenses	13	0.01	0.01	0.01	0.02	0.01	0.01	0.01	0.02
Miscellaneous / Other	15	0.18	0.09	0.02	0.36	0.18	0.09	0.02	0.36
Fixed									
Real Estate Taxes	26	1.53	1.49	1.17	1.81	0.53	1.49	1.17	1.84
Personal Property Tax	6	0.08	0.00	0.00	0.22	0.08	0.00	0.00	0.22
Other Tax									
Building Insurance	25	0.14	0.13	0.11	0.15	0.14	0.13	0.11	0.15
License / Fees / Permits	12	0.04	0.03	0.01	0.07	0.04	0.03	0.01	0.07
Directly Expensed Leasing									
Payroll									
Commissions / Fees	9	1.15	0.81	0.60	1.47	1.15	0.82	0.60	1.47
Advertising / Promotion	12	0.08	0.04	0.01	0.09	0.08	0.04	0.01	0.09
Professional / Fees	12	0.04	0.02	0.01	0.04	0.04	0.02	0.01	0.04
Tenant Improvements	15	3.02	1.95	0.43	4.47	3.03	1.95	0.43	4.54

Figure 9.3d Atlanta income and expense detail (*continued*)

Other Leasing Costs	13	0.61	0.05	0.01	0.07	0.61	0.05	0.01	0.07
Amortized Leasing									
Commissions / Fees	9	0.58	0.52	0.15	0.85	0.58	0.53	0.15	0.85
Tenant Improvements	9	0.73	0.41	0.14	1.27	0.74	0.41	0.14	1.27
Other Leasing Costs									
Parking									
In-house									
Contract									
Snow Removal									
Shuttle									

Figure 9.3e Atlanta income and expense detail (*continued*)

Fixed Expenses

BOMA data is used to budget all fixed expenses except for reserves for replacement. The amount for reserves for replacement is automatically transferred to this worksheet from the CAP-X worksheet computation. Other cautions:

- Real estate tax—check local tax rates for specific site locations. BOMA data covers a general area in each city and rates can vary widely.
- Insurance—check rates with an insurance agent to include insurance for hurricane, tornado, and seismic activity if desired.
- Property management—if for a commercial building, add sufficient budget from BOMA data to cover what they call "leasing expenses."

And last, the author does not include tenant alteration expense in the budget because normally that is included as capital improvement/investment in the next lease for the space. Some lenders require that a tenant improvement (TI) budget reserve be established to refurbish space for a new tenant when the existing lease expires. If so, include this in the CAP-X budget.

Lessor's Annual Cost Statement

The third worksheet in the annualcost.xls set is called "1217." This name is derived from use of GSA's Form 1217, Lessor's Annual Cost Statement.

GSA Form 1217 is used by all owners and developers who submit bids for space to GSA for leasing. Figure 9.4 illustrates this form for the sample project.

Completion of this form requires the user to perform just a few simple steps:

1. At the top of the 1217 form, fill in Box 3, Net Rentable Area.
2. Next, fill in Box 3a, Net Rentable Area for the entire building. Most of the time this is the same as the figure entered in Box 3.
3. Lastly, fill in Box 3b, Net Rentable Area leased by the government (or other tenant).

GENERAL SERVICES ADMINISTRATION PUBLIC BUILDINGS SERVICE	1. LEASE OR BID INVITATION NO.		2. STATEMENT DATE
LESSOR'S ANNUAL COST STATEMENT IMPORTANT **Read "Instructions" on reverse of form.**	3. NET RENTABLE AREA (Sq ft.) 203,850	3A. ENTIRE BUILDING 203,850	3B. LEASED BY GOV'T

4. BUILDING NAME AND ADDRESS   (No., street, city, State and zip code no.)

Office Building Headquarters   Atlanta, Georgia

SECTION I ▯ ESTIMATED ANNUAL COST OF SERVICES AND UTILITIES FURNISHED BY LESSOR AS PART OF RENTAL CONSIDERATION

SERVICES AND UTILITIES	LESSOR'S ANNUAL COST FOR		FOR GOVERNMENT
	(a) ENTIRE BUILDING	(b) GOV'T LEASED AREA	USE ONLY (c)
A. CLEANING, JANITOR AND/OR CHAR SERVICE 5. SALARIES	161,042	–	
6. SUPPLIES (Wax, cleaners, cloths, etc.)	22,424	–	
7. CONTRACT SERVICES (Window washing, waste and snow removal)	18,347	–	
B. HEATING 8. SALARIES			
9. FUEL ("X" ONE) | |OIL| |GAS| |COAL| |ELECTRIC	190,575	–	
10. SYSTEM MAINTENANCE AND REPAIR	16,308	–	
C. ELECTRICAL 11. CURRENT FOR LIGHT AND POWER (including elevators)	129,346	–	
12. REPLACEMENT OF BULBS, TUBES, STARTERS	6,116	–	
13. POWER FOR SPECIAL EQUIPMENT	–	–	
14. SYSTEM MAINTENANCE AND REPAIR (Ballasts, fixtures, etc.)	12,231	–	
D. PLUMBING 15. WATER (For all purposes) (Include sewer charges)	34,655	–	
16. SUPPLIES (Soap, towels, tissues not in 6 above)	–	–	
17. SYSTEM MAINTENANCE AND REPAIR	6,116	–	
E. AIR CONDITIONING 18. UTILITIES (Include electricity, if not in C 11)	275,576	–	
19. SYSTEM MAINTENANCE AND REPAIR	34,655	–	
F. ELEVATORS 20. SALARIES (Operators, starters, etc.)			
21. SYSTEM MAINTENANCE AND REPAIR	20,385	–	
G. MISCELLANEOUS 22. BUILDING ENGINEER AND/OR MANAGER	75,425	–	
23. SECURITY (Watchman, guards, not janitors)	81,540	–	
24. SOCIAL SECURITY TAX AND WORKMEN'S COMPENSATION INSURANCE	12,231	–	
25. LAWN AND LANDSCAPING MAINTENANCE	44,847	–	
26. OTHER Emergency Generator	6,116	–	
27. Total	$ 1,147,931	$ –	$

SECTION II ▯ ESTIMATED ANNUAL COSTS OF OWNERSHIP EXCLUSIVE OF CAPITAL CHARGES

28. REAL ESTATE TAXES	311,891	–	
29. INSURANCE (Hazard, liability, etc.)	28,539	–	
30. BUILDING MAINTENANCE AND RESERVES FOR REPLACEMENT	240,104	–	
31. MANAGEMENT	203,850	–	
32. Total	$ 784,383	$ –	$

LESSOR'S CERTIFICATION ▯ The amounts entered in Columns (a) and (b) represent my best estimate as to the annual costs of services, utilities and ownership.	33. SIGNATURE OF ☐ OWNER ☐ LEGAL AGENT

SECTION III ▯ APPROVAL OF STATEMENT BY AUTHORIZED GOVERNMENT REPRESENTATIVES

The undersigned certify that the amount shown in Item 27(c) represents the reasonable value of the services and utilities which amount may be properly deducted in determining net rent.

TYPED NAME AND TITLE	SIGNATURE	DATE
34A.	34B	34C.
35A.	35B.	35C.

GSA Form 1217

Figure 9.4 GSA Form 1217 for sample project

The computer will bring forward all costs from the O&M work-sheet, place them in the proper GSA categories, allocate them on a percentage of occupancy basis between columns (a) and (b) on the form, and then total the columns.

Rent Computation

The rent computation format is the final worksheet in the softcost. xls series. It is marked "RENT" in that series.

To calculate rent all the reader now must do is fill in five shaded squares in the format:

1. Enter the rented area.
2. Enter the operating expenses (line 27 from GSA Form 1217).
3. Enter the fixed expenses (line 32 from GSA Form 1217).
4. Enter the annual interest rate (in whole number format) for the total loan amount. (The total loan amount is brought forward automatically from previous worksheets.)
5. Enter the amortization period for the loan (in years).

The program computes the total rent per square foot as illustrated in Figure 9.5.

Rent Computation

Project: Office Building Headquarters   Date _____
Location: Atlanta, Georgia

Rented Area		203,850	SF
Operating Expenses	$	1,147,931	
Fixed Expenses	$	784,383	

Debt Service Calculation

Total Loan Amount	$	60,508,210	
Annual Interest rate		5.00	%
Amortization period		30.00	Years

Capital payment	$	324,821.16	per month	$ 1.59 per sf
O&M required	$	161,026.17	per month	$ 0.14 per sf
Total	$	485,847.32	per month	$ 1.73 per sf
Total	$	5,830,167.87	per year	**$ 28.60** per sf

Figure 9.5 Rent computation for sample project

Evaluating Rent

Once you have computed the rental rate you now must determine if it is worth what the market will support. If the rent is too high you can go back through the models to adjust quantities of space, configuration, design parameters, system types and materials, soft costs, schedule, financing, and operating costs to see if you can lower the rent required and still achieve the project objectives.

If the rental rate comes out below market, then you have a potential winner on your hands. In other words, it is worth more than it cost!

For our sample project in Atlanta, notice that BOMA data indicates the maximum rent there is $23.92 per square foot (Figure 9.3a) and the author's calculation came out at $28.60. That means there is some work to do to reduce the calculated rent to make the project more viable.

Note

1. *BOMA Experience Exchange Report (EER)*, Washington, D.C., www.boma.org (accessed 15 August 2013).

10 Using Your Budget

Need for Cost Control

Now that you have created a project budget out of thin air to meet a stated procurement requirement it is your main lifeline to success should you be announced the winner. For a stated rent "until death do us part" you have agreed to operate and maintain a building for its life that does not even exist.

Developers have a critical need for cost control if they have justified the project budget to a lender or investor promising to give them a certain return on investment. What is the consequence if you fail to achieve the budget? Figure 10.1 illustrates some of the economic consequences that are possible. It provides the economic summary of a project that was budgeted to provide a return-on-equity investment (ROI) of 18.1 percent.

Several scenarios are presented to show what would happen if budgeted costs increased:

- First, assume project construction cost rose 10 percent over that originally planned. This added increase in initial cost reduces the ROI to 12.1 percent.
- In the second situation, assume that the designed facility operating costs are 10 percent higher than planned. If this were the case, ROI would become 13.4 percent.
- In the third situation, both the construction costs and the operating costs are assumed to be 10 percent higher than originally planned. The net result is an ROI of only 9.6 percent.

In all three cases, the resulting ROI is less than the owner would have accepted during the planning phase and certainly represents a poor return on investment for the risk and effort involved.

	As Planned (Budgeted)	Construction Costs +10%	Operation Costs +10%	Construction & Operation Costs +10%	Economic Studies Construction −5% Operation −5%
Total Construction Cost	34,757,000	38,233,000	34,757,000	38,233,000	33,020,000
Indirect Costs (includes economics)	9,249,000	9,711,000	9,247,000	9,247,000	9,062,000
Land Cost	4,480,000	4,480,000	4,480,000	4,480,000	4,480,000
Total Project Cost	48,486,000	52,424,000	48,486,000	52,424,000	46,562,000
Less Mortgage Loan*	40,583,000	40,583,000	38,384,000	38,384,000	41,686,000
Equity Investment Required	7,903,000	11,841,000	10,102,000	14,040,000	4,876,000
Gross Income	8,850,000	8,850,000	8,850,000	8,850,000	8,850,000
Operating Cost	3,110,000	3,110,000	3,241,000	3,241,000	2,954,000
Net Income*	5,740,000	5,740,000	5,429,000	5,429,000	5,896,000
Less Mortgage Payment (debt service)	4,305,000	4,305,000	4,072,000	4,072,000	4,422,000
Before Tax Stabilized Cash Flow	1,435,000	1,435,000	1,357,000	1,357,000	1,474,000
Return on Equity Investment	18.10%	12.10%	13.40%	9.60%	30.20%

*Loan amount determined by 75% of capitalized (@ 10%) net income
0.75 x $5,750,000 x 9.427 (PWA) = $40,538,000
0.75 x $5,429,000 x 9.427 (PWA) = $38,384,000
0.75 x $5,896,000 x 9.427 (PWA) = $41,686,000

Figure 10.1 Economic impacts of cost change

This figure also illustrates the need to control follow-on costs. Although operation costs are not the main subject of this book, there are texts that treat this area.[1]

Considering how easily 10 percent changes can "creep" into a project the importance of an effective cost control effort becomes apparent. The entire financial feasibility of a project can be drastically altered long before anything is in the ground.

• In the final situation, suppose both planned initial cost and ownership cost were reduced by a mere 5 percent. The result is an ROI of 30.2 percent.

This result can easily be achieved by monitoring project costs throughout design and applying both value engineering and lifecycle costing techniques. First, however, budget scope must be adequately defined.

Using the Product of this Book

Once project scope has been defined as described in this book it provides a most useful tool for the remaining facility acquisition process and ownership.

If the next step is hiring a professional consulting firm to provide design work for construction, the scope documents should be a condition of that contract.

If the next step is the direct procurement of a facility through the use of a developer or design-build contract, the scope documents should be incorporated as a basis for that procurement. When scope is properly defined, it provides the following:

1. a clear understanding between those who have proposed the scope and the owner, who has accepted the scope;
2. a document to be used by the designer or developer that indicates clearly what is to be provided in terms of technical content, quality levels, schedule, and cost;
3. a baseline in sufficient detail for use in controlling delivered work to permit proper evaluation of any subsequent changes or deviations for decision making;
4. a tool to permit subsequent evaluation of the accuracy and performance of various organizations involved in the acquisition process by comparing the intended result against the actual experience;
5. a definitive cost estimate from which to control the design, procurement, construction, and start-up of the owner's facility.

Controlling Design Work

If the objectives of this form of budgeting are to be met, then the proof will be in the resulting design procurement. It is one thing to set realistic budgets at a conceived level of requirements for a specific project and another thing to communicate those requirements to all members of the building team involved in the procurement.

This is made more difficult to the extent that an owner, like a major corporation that builds many facilities, custom tailors its requirements on a project by project basis. Therefore, project scope control as a means of cost control becomes extremely important.

A-E Design

Cost control to the budget must start in the initial contract for design services. The contract should be written to include the following features:

- The approved maximum ECC (estimated construction cost) component of the maximum ETPC should constitute the budget goal.

The figure used for A-E contract negotiations should be the maximum ECCA (estimated cost of construction at award), adjusted to the anticipated date of construction contract award. Maximum ECCA is the maximum approved ECC, less contingencies and reservations.

- Incorporate in the contract scope of service the approved project cost plan that sets forth the costs for each building element at UNIFORMAT Level 3 and indicates the assumptions upon which the cost figures are based.

Provide the full design program developed to support the cost plan. The estimate developed for each system and each aspect of the total project cost should constitute target cost and quality levels to be obtained.

- Indicate that the A-E is free to design a building that differs from the configuration used in determining the budget figure; however, cost and quality targets must not be exceeded.

The project cost plan should be rigorously enforced through the design process. The A-E should continue to be required, by contract, to redesign the project at the A-E's own expense if the bid cost exceeds budget by 5 percent, providing the owner did not impose extenuating circumstances, such as adding scope to the project or changing criteria.

- Require that the A-E deliver an independently prepared cost estimate at each stage of the design.

The estimate should be prepared on the basis of the UNIFORMAT breakdown code of accounts and summarized at Level 3 to compare with the budget on both a parameter and cost per gross square foot basis.

- If the estimates submitted by the A-E exceed budget, the A-E should submit a list of cost savings, trade-offs, or other reductions to bring cost under control.

All A-E cost savings should be subject to owner approval, and all approved changes to delivered design work should be implemented by the A-E without additional cost to the owner.

Design-Build Procurement

With this type of procurement, cost control is largely the contractor's responsibility. It is assumed that the owner used the full design program as a contract requirement and signed a contract within budget.

The owner must require the developer to submit the required design information for checking against the quantity and quality standards delineated in the cost plan.

Capturing Cost Experience

One learns from experience, and so it is with budgeting. Each time parameter estimating is used to prepare a budget should be a learning experience that will improve the performance for next time.

The developer can check the prediction of system quantity and quality against actual final design work. However, one needs actual cost data on a systems basis to validate and/or develop parameter costs. This data can come from contractor quotations requested in advance.

All contractors want progress payments for their work. The owner can decide the basis for granting those payments. Simply place a blank UNIFORMAT breakdown, with units of measure, in the contract to be filled out by the contractor to be used as the basis for progress payments. A sample payment schedule format is published by AIA.[2]

Have the contractor fill in the quantities and unit prices and make the extensions. Use your own A-E prepared UNIFORMAT estimate as a basis for approval. Check both cost and quantities for reasonableness and guard against front-end loading of cost.

Summary

Many years ago the author prepared and submitted a budget through top management to Congress for a new federal office building. The cost of the building calculated to a price of $75.00/gsf, which at the time seemed quite high. Management questioned the cost saying that data from the private sector indicated that a comparable commercial office building could be constructed for $60.00/gsf.

An explanation of the difference was demanded. This was an extremely difficult task. How can anyone identify the differences between two supposedly similar types of construction when all one knows is cost per square foot? At that moment, realization set in that cost control could not be achieved and was never achieved for projects budgeted on a cost per square foot basis.

To "control" something one has to review future produced work against the original baseline in order to identify the variances so that the project manager can decide on solutions to the variances. Gross square foot as the budget baseline does not give much to use when it is the sole check against future design work. To control cost you must control the quality of construction being designed and the true program, technical content, and criteria to which the project is being designed. Therefore, it is these things that must be used as the basis for project budgeting.

Until now, the development of design parameters associated with a design program, design criteria, and a design concept was not practical. It took longer and cost more than could be afforded for producing a budget. However, with the use of the design programming professional services in the last decade, the establishment of the UNIFORMAT system of estimating, increased use of the computer, and the availability of commercial cost data sources providing data support for UNIFORMAT, program needs can be specified and budgeted based on selected design criteria.

It is practical and necessary to calculate basic parameter quantities based on codes and criteria prior to beginning design. Simulating the building in such a manner permits one to budget based on a specific planned level of quality, performance, content, and capability. This, then, can be reviewed during the budget cycle and changed by management as desired to meet their policy decisions. Once budgeted in this manner, the approved criteria and systems can be communicated to the architect and contractor as a basis for design and construction and can be used to control the design and its cost.

Notes

1. Dell'Isola, A.J. and Kirk, S.J., *Life Cycle Costing for Design Professionals*, 1981, New York: McGraw Hill.
2. Section B-5, Design and Construction Cost Management, *Architect's Handbook of Professional Practice*, vol. I, 1985, Washington, D.C.: AIA.

Appendix A
"Zipped" Format Index

The following files accompanying this text are available for free download from the publisher's website: www.routledge.com/97811 38016156

File Name:	Use
mspace.xls	Space model preparation
mconfig.xls	Configuration model
mhvac.xls	HVAC model computations
mplumb.xls	Plumbing model
melect.xls	Electrical model
P-89mod.pdf	Weather data for use in mhvac.xls
costtemplate.xls	Construction cost estimate format
data2014.xls	Construction cost database for year 2014*
softcost.xls	Workbook to calculate all project costs and rent
annualcost.xls	Workbook to calculate building operating costs
Estimating Instructions.doc	
Maintaining the Data Base.doc	

* The file provides for the ability to escalate costs in the database. It is suggested that you do this annually in November of each year and do not forget to change the file name year and resave it.

Appendix B
Uniformat Code of Accounts

The author uses the original version of UNIFORMAT in this text because it is much more detailed in describing what is contained in each code of accounts. If you wish to format your estimate in the revised version of UNIFORMAT you can modify costtemplate. xls accordingly as follows.

(Original Version)	(Revised Version II)
01 Foundations	A. Substructure
02 Substructure	(incl)
03 Superstructure	B. Superstructure
04 Exterior Closure	(incl)
05 Roofing	(incl)
06 Interior Construction	C. Interior Construction
07 Conveying Systems	(incl)
08 Mechanical	D. Services
081 Plumbing	(incl)
082 HVAC	(incl)
083 Fire Protection	(incl)
084 Special Mechanical	(incl)
09 Electrical	(incl)
091 Service and Distribution	(incl)
092 Lighting and Power	(incl)
093 Special Electrical Systems	(incl)
10 General Conditions and Profit	(incl)
11 Equipment	E. Equipment
112 Furnishings	F. Special Construction
12 Site Work	G. Site Work

01   Foundations

011   Standard Foundations

0011   Wall Foundations
01111   Wall Footings
01112   Foundation Walls and Pilasters
01113   Excavating and Backfilling

0112   Column Foundations and Pile Caps
01121   Column Footings
01122   Pile Caps
01123   Column Piers and Base
01124   Excavating and Backfilling

012   Special Foundation Conditions

0121   Pile Foundations
01211   Mobilization–Demobilization
01212   Pile Tests
01213   Piles

0122   Caissons
01221   Open Caissons
01222   Caisson Accessories
01223   Special Caissons

0123   Underpinning
01231   Temporary Shoring to Structure
01232   Excavating
01233   Sheeting and Shoring to Structure
01234   Backfilling
01235   Concreting
01236   Formwork
01237   Steel Bar Reinforcing
01238   Cutoff Projecting Footings
01239   Grouting and Dry Packing

0124   Dewatering
01241   Pumping
01242   Well-Point
01243   Gravity Drainage

022   Basement Excavation

0221   Excavation for Basement
02211   Excavating
02212   Waste Material Disposal

0222   Structure Backfill and Compaction
02221   Structure Backfill with Excavated Material
02222   Borrow Backfill

0223   Shoring
02231   Sheeting and Shoring
02232   Tiebacks and Anchors
02233   Slurry Walls

023   Basement Walls

0231   Basement Wall Construction
02311   Basement Walls
02312   Pilasters
02313   Expansion and Construction Joints

0232   Moisture Protection
02321   Dampproofing
02322   Waterproofing

0233   Basement Wall Insulation

03   Superstructure

031   Floor Construction

0311   Suspended Basement Floor Construction
03111   Structural Frame
03112   Interior Structural Walls
03113   Floor Slabs and Decks
03114   Inclined and Stepped Floors
03115   Expansion and Contraction Joints

0312   Upper Floor Construction
03121   Structural Frame
03122   Interior Structural Walls

03123   Floor Slabs and Decks
03124   Inclined and Stepped Floors
03125   Expansion and Contraction Joints

0313   *Balcony Construction*
03131   Supported Balconies
03132   Cantilevered Balconies

0314   *Ramps*
03141   Pedestrian Ramps
03142   Vehicle Ramps

0315   *Special Floor Construction*
03151   Catwalks
03152   Space Frames
03153   Cable Supported Floors

032   *Roof Construction*

0321   *Flat Roof Construction*
03211   Structural Frame
03212   Interior Structural Walls
03213   Roof Slabs and Decks
03214   Expansion and Contraction Joints

0322   *Pitched Roof Construction*
03221   Frame and Trusses
03222   Roof Decking and Sheathing

0323   *Canopies*
03231   Supported Canopies
03232   Cantilevered Canopies

0324   *Special Roof Systems*
03241   Concrete Shells/Domes
03242   Hyperbolic Parabaloid Roofs
03243   Space Frames
03244   Barrel Vault Roofs
03245   Saw Tooth Roofs
03246   Cable Supported Roofs
03247   Air Supported Structures

033   *Stair Construction*

0331   *Stair Structure*
03311   Regular Stairs
03312   Curved Stairs
03313   Spiral Stairs
03314   Exterior Fire Escapes
03315   Steps in Suspended Slabs

04   Exterior Closure

041   *Exterior Walls*

0411   *Exterior Wall Construction*
04111   Exterior Skin
04112   Insulation and Vapor Barriers
04113   Interior Skin
04114   Parapets
04115   Dampproof Courses
04116   Finish to Exposed Structure
04117   Expansion Joints
04118   Cornerstones

0412   *Exterior Louvers and Screens*
04121   Exterior Louvers
04122   Decorative Grilles and Screens
04123   Exterior Vents

0413   *Sun Control Devices (Exterior)*
04131   Projecting Sun Screens
04132   Awnings
04133   Exterior Shutters/Blinds

0414   *Balcony Walls and Handrails*
04141   Balcony Walls
04142   Balcony Railings
04143   Balcony Handrails
04144   Balcony Dividing Walls

0415   *Exterior Soffits*
04151   Building Soffits

05022   Waterproof Membranes Below Paving
05023   Slatted Roof Decks and Walkways

0503   Roof Insulation and Fill
05031   Roof Vapor Barriers
05032   Roof and Deck Insulation
05033   Roof Fill

0504   Flashings and Trim
05041   Flashings
05042   Gravel Stops
05043   Fascia and Eaves
05044   Gutters and Downspouts
05045   Miscellaneous Roofing Specialties

0505   Roof Openings
05051   Glazed Roof Openings
05052   Hatches
05053   Gravity Roof Ventilators

06   Interior Construction

061   Partitions

0611   Fixed Partitions
06111   Solid Partitions
06112   Glazed Partitions
06113   Mesh Partitions

0612   Demountable Partitions
06121   Full Height Demountable Partitions
06122   Bank Height Demountable Partitions

0613   Retractable Partitions
06131   Accordion Folding Partitions
06132   Folding Leaf Partitions
06133   Coiling Partitions

0614   Compartments/Cubicles
06141   Toilet Partitions

06142   Shower and Dressing Compartments
06143   Hospital Cubicles

0615   *Interior Balustrades and Screens*
06151   Stair Balustrades
06152   Balustrades at Floor Openings
06153   Interior Grilles and Decorative Screens

0616   *Interior Doors and Frames*
06161   Interior Doors
06162   Interior Door Frames
06163   Interior Door Hardware
06164   Interior Door Wall Opening Element
06165   Interior Door Sidelights and Transoms
06166   Interior Door Painting and Staining
06167   Hatches/Access Doors

0617   *Interior Storefronts*
06171   Framing
06172   Panels and Bulkheads
06173   Doors and Hardware
06174   Glazing
06175   Rolling Grilles and Folding Closures

062   *Interior Finishes*

0621   *Wall Finishes*
06211   Wall Finishes to Inside Exterior Wall
06212   Wall Finishes to Interior Walls
06213   Column Finishes

0622   *Flooring*
06221   Screens and Toppings
06222   Floor Finishes
06223   Bases, Curbs, and Trim
06224   Stair Finish
06225   Access Flooring (Pedestal Floors)

0623   *Ceiling Finishes*
06231   Ceiling Finishes Applied to Structure
06232   Suspended Ceilings

06233   Special Ceilings
06234   Stair Soffits
06235   Expansion Joint Covers

063   Specialties

0631   General Specialties
06311   Chalk and Tackboards
06312   Identifying Devices
06313   Lockers
06314   Toilet/Bath Accessories
06315   Storage Shelving
06316   Misc. Metalwork
06317   Misc. Specialties

0632   Built-in Fittings
06321   Counters and Vanities
06322   Kitchen Cabinets
06323   Closets
06324   Miscellaneous Built-in Cabinetwork

07   Conveying Systems

0701   Elevators
07011   Passenger Elevators
07012   Freight Elevators

0702   Moving Stairs and Walks
07021   Escalators
07022   Moving Walks

0703   Dumbwaiters
07031   Hand-Operated Dumbwaiters
07032   Electric-Operated Dumbwaiters

0704   Pneumatic Tube System
07041   Pneumatic Message Tube Systems
07042   Pneumatic Trash Tube Systems
07043   Pneumatic Linen Tube Systems

0705   *Other Conveying Systems*
07051   Lifts
07052   Hoists and Cranes
07053   Conveyors
07054   Chutes
07055   Turntables

0706   *General Construction Items*
07061   Hoistway Beams
07062   Hydraulic Elevator Shaft Drilling
07063   Miscellaneous Metals
07064   Lintels to Openings
07065   Concrete Work
07066   Masonry Work
07067   Painting

08   *Mechanical*

081   *Plumbing*

0811   *Domestic Water Supply System*
08111   Cold Water Service
08112   Hot Water Service
08113   Domestic Water Supply Equipment

0812   *Sanitary Waste and Vent System*
08121   Waste Piping and Fittings
08122   Vent Piping and Fittings
08123   Floor Drains
08124   Sanitary Waste Equipment
08125   Thermal Pipe Insulation

0813   *Rainwater Drainage System*
08131   Pipe and Fittings
08132   Roof Drains
08133   Rainwater Drainage Equipment
08134   Thermal Pipe Insulation

0814   *Plumbing Fixtures*
08141   Bath Tubs

08142 Bidets
08143 Kitchen Sinks
08144 Laundry Sinks and Trays
08145 Lavatories
08146 Mop Sinks
08147 Service Sinks
08148 Showers
08149 Urinals
081410 Water Closets
081411 Wash Fountains
081412 Drinking Fountains and Coolers

082   *HVAC*

0821   *Energy Supply*
08211 Oil Supply System
08212 Gas Supply System
08213 Coal Supply System
08214 Steam Supply System
08215 Solar Energy Supply System
08216 Wind Energy Supply System

0822   *Heat Generating System*
08221 Steam Boilers
08222 Hot Water Boilers
08223 Furnaces
08224 Boiler Room Piping and Specialties
08225 Auxiliary Equipment
08226 Equipment Thermal Insulation

0823   *Cooling Generating Systems*
08231 Chilled Water Systems
08232 Direct Expansion Systems

0824   *Distribution Systems*
08241 Air Distribution
08242 Exhaust Ventilation Systems
08243 Steam Distribution
08244 Hot and Chilled Water Distribution
08245 Change-Over Distribution Systems
08246 Glycol Heating Distribution System

0825   *Terminal/Package Units*
08251   Terminal Units
08252   Packaged Units

0826   *Controls and Instrumentation*
08261   Air-Conditioning System
08262   Energy Supply System
08263   Heat-Generating System
08264   Cooling-Generating System
08265   Special Mechanical Systems
08266   Instrument Panels
08267   Instrument Air Compressor
08268   Gas Purging System

0827   *Systems Testing and Balancing*
08271   Water Side Testing and Balancing
08272   Air Side Testing and Balancing

083   *Fire Protection*

0831   *Water Supply (Fire Protection)*
08311   Water Connection
08312   Pipe and Fittings
08313   Valves

0832   *Sprinklers*
08321   Wet Sprinkler System
08322   Dry Sprinkler System
08323   Standpipe Systems
08331   Standpipe Equipment
08332   Fire Hose Equipment
08333   Pumping Equipment

0834   *Fire Extinguishers*
08341   Hand-Held Fire Extinguishers
08342   Wheeled Cart Fire Extinguishers
08343   Fire Extinguisher Cabinet

084   *Special Mechanical Systems*

0841   *Special Plumbing Systems*
08411   Special Piping Systems

08412   Acid Waste Systems
08413   Interceptors
08414   Pool Equipment
08415   Special Plumbing Fixtures

0842   Special Fire Protection Systems
08421   Carbon Dioxide Extinguishing Equipment
08422   Foam-Generating Equipment
08423   Halon System Equipment
08424   Hood and Duct Fire Protection

0843   Miscellaneous Special Systems and Devices
08431   Special Cooling Systems and Devices
08432   Process Heating
08433   Storage Cells and Devices
08434   Dust and Fume Collectors
08435   Deodorizing Equipment
08436   Carbon Monoxide Equipment
08437   Sound Attenuating Equipment
08438   Special Waste Treatment Devices

09   Electrical

091   Service and Distribution

0911   High Tension Service and Distribution
09111   High Tension System Monitoring
09112   High Tension System Equipment
09113   High Tension System Distribution

0912   Low Tension Service and Distribution
09121   Low Tension System Monitoring
09122   Low Tension System Equipment
09123   Low Tension System Distribution

092   Lighting and Power

0921   Branch Wiring
09211   Wiring Circuits
09212   Branch Wiring Devices

0922   *Lighting Equipment*
09221 Fluorescent Interior Lighting Fixtures
09222 Incandescent Interior Lighting Fixtures
09223 Other Lighting Fixtures and Equipment

093   *Special Electrical Systems*

0931   *Communications and Alarm Systems*
09311 Public Address Systems
09312 Central Music Systems
09313 Intercommunication Systems
09314 Paging Systems
09315 Utility Telephone Systems
09316 Nurses' Call System
09317 In–Out Registers
09318 Bell Systems
09319 Television Systems
093110 Clock and Program Systems
093112 Burglar Alarm Systems
093113 Other Systems

0932   *Grounding Systems*
09321 Lightning Protection
09322 Building Ground System
09323 Special Grounding Systems

0933   *Emergency Light and Power*
09331 Emergency Generator Systems
09332 Emergency Battery Systems
09333 Other Emergency Light and Power Systems

0934   *Electric Heating*
09341 Heating Equipment
09342 Control Devices
09343 Branch Wiring
09344 Other Heating Systems

0935   *Floor Raceway Systems*
09351 Standard Under-Floor Duct Systems
09352 Header (Feeder) Duct

09353 Industrial (Square) Duct
09354 Trench Duct
09355 Wiring Devices and Accessories

0936   Other Special Systems and Devices
09361 Special Lighting System
09362 Special Protective Systems and Devices
09363 Special Electronic Controls

0937   General Construction Items
09371 Cutting and Patching
09372 Trenching and Backfill
09373 Painting
09374 Equipment Installation Items

10   General Conditions and Profit

1001   Mobilization and Initial Expenses
10011 Mobilization
10012 Permits and Fees
10013 Insurance and Bonds

1002   Site Overhead
10021 Site Supervisory and Emergency Staff
10022 Labor Onsite Costs
10023 Sales and Use Taxes
10024 Construction Equipment
10025 Site Office Operating Costs
10026 Temporary Facilities
10027 Site Protection Security
10028 Cleanup
10029 Inspection and Testing
100210 Winter Conditions
100211 Miscellaneous Site Overhead

1003   Demobilization
10031 Temporary Enclosures (Removal)
10032 Temporary Buildings (Removal)
10033 Temporary Services (Removal)
10034 Equipment Demobilization
10035 Final Clean-up

10036   Repairing Sidewalks and Streets
10037   Punch List and Warranties
10038   Maintenance Manuals and As-Built Drawings
10039   Staff Relocation Costs
100310 Opening Ceremonies

1004   *Main Office Expense and Profit*
10041   Main Office Expense
10042   Profit

11 *Equipment*

111   *Fixed and Movable Equipment*

1111   *Built-in Maintenance Equipment*
11111   Window Washing Equipment
11112   Vacuum Cleaning System

1112   *Checkroom Equipment*
11121   Manual Checkroom Equipment
11122   Automatic Storage and Retrieval Checkroom Equipment

1113   *Food Service Equipment*
11131   Refrigeration Cases
11132   Insulated Rooms
11133   Storage Units
11134   Cooking Equipment
11135   Food Preparation Machines
11136   Food Serving Units
11137   Washing Units and Conveyors

1114   *Vending Equipment*
11141   Hot Drink Vending Unit
11142   Cold Drink Vending Unit
11143   Hot Food Vending Unit
11144   Cold Food Vending Unit
11145   Cigarette Vending Unit
11146   Condiment Unit and Counter
11147   Refuse Unit
11148   Coin Changer
11149   Microwave Oven
111410 Bases for Unit

1115   Waste Handling Equipment
11151   Waste Compactors
11152   Incinerators
11153   Waste Storage Containers
11154   Pulping Machines and System

1116   Loading Dock Equipment
11161   Dock Levellers
11162   Levelling Platforms
11163   Dock Bumpers
11164   Dock Seals and Shelters

1117   Parking Equipment
11171   Parking Bumpers and Guard Rails
11172   Parking Control Equipment

1118   Detention Equipment
11181   Cell and Corridor Construction
11182   Cell Accessories
11183   Courtroom Security Devices
11184   Detention Screens

1119   Postal Equipment
11191   Mail Boxes
11192   Post Office Equipment

11110   Other Specialized Equipment
111101   Darkroom Equipment
111102   Educational Equipment
111103   Athletic Equipment
111104   Laboratory Equipment
111105   Laundry Equipment
111106   Library Equipment
111107   Medical Equipment
111108   Mortuary Equipment
111109   Residential Equipment
111110   Auditorium and Stage Equipment
111111   Misc. Specialized Equipment

112   Furnishings

1121   Artwork
11211   Bases and General Contract Work for Artwork and Sculpture

12112   Tree Removal
12113   Selective Thinning
12114   Tree Pruning

1212   Demolition
12121   Building Demolition
12122   Site Demolition
12123   Relocations

1213   Site Earthwork
12131   Site Grading
12132   Site Excavating
12133   Borrow Fill
12134   Soil Stabilization
12135   Soil Treatment
12136   Site Dewatering
12137   Site Shoring

122   Site Improvements

1221   Parking Lots
12211   Parking Lot Paving and Surfacing
12212   Curbs, Rails, and Barriers
12213   Parking Booths and Equipment

1222   Roads, Walks, and Terraces
12221   Roads
12222   Walks
12223   Terraces and Plazas

1223   Site Development
12231   Fences and Gates
12232   Walls
12233   Signs
12234   Site Furnishings
12235   Fountains, Pools, and Watercourses
12236   Playing Field and Sports Facilities
12237   Flagpoles
12238   Miscellaneous Structures

1224   Landscaping
12241   Fine Grading and Soil Preparation

12242   Top Soil and Planting Beds
12243   Seeding and Sodding
12244   Planting
12245   Planters
12246   Special Landscape Feature

123   *Site Utilities*

1231   *Water Supply and Distribution*
12311   Potable Water Systems
12312   Fire Protection Systems
12313   Process Water Systems
12314   Irrigation Systems

1232   *Drainage and Sewerage Systems*
12321   Storm Drainage
12322   Sanitary Sewer
12323   Process and Acid Waste Systems
12324   Combined Drainage and Sewerage Systems

1233   *Heating and Cooling Systems*
12331   Heating System
12332   Cooling Systems

1234   *Gas Distribution Systems*
12341   Natural Gas Systems
12342   Other Gas Systems

1235   *Electric Distribution and Lighting Systems*
12351   Overhead Power Service
12352   Underground Services
12353   Exterior Yard and Road Lighting
12354   Exterior Flood Lighting
12355   Exterior Lighting Controls
12356   Exterior Sign Lighting

1236   *Snow Melting Systems*
12361   Piped Snow Melting Systems
12362   Electrical Snow Melting Systems

1237   *Service Tunnels*
12371   Excavating and Backfilling

12372   Constructed Service Tunnel
12373   Prefabricated Service Tunnels
12374   Moisture Protection
12375   Insulation
12376   Miscellaneous Items

124   Off-Site Work

1241   Railroad Work

1242   Marine Work

1243   Tunneling

1244   Other Off-Site Work

Index

Page references in *italics* indicate figures.